FAWN ISLAND

FAWN ISLAND

*Written and Illustrated
by Douglas Wood*

University of Minnesota Press
Minneapolis
London

Published by the University of Minnesota Press
111 Third Avenue South, Suite 290
Minneapolis, MN 55401-2520
http://www.upress.umn.edu

Library of Congress Cataloging-in-Publication Data

Wood, Douglas,
    Fawn Island / written and illustrated by Douglas Wood.
      p.   cm.

    ISBN 978-0-8166-3176-6 (pb)

    1. Rainy Lake Region (Minn. and Ont.)—Description and travel. 2. Rainy Lake Region (Minn. and Ont.)—History. 3. Rainy Lake Region (Minn. and Ont.)—Pictorial works. 4. Natural history—Rainy Lake Region (Minn. and Ont.). 5. Islands—Rainy Lake Region (Minn. and Ont.). 6. Wood, Douglas, 1951– I. Title.
    F612.R18 W66 2001
    917.76'79—dc21

                      00-010660

Printed in the United States of America on acid-free paper

The University of Minnesota is an equal-opportunity educator and employer.

23 22 21 20 19 18 17   10 9 8 7 6 5 4 3 2 1

Now one autumn morning when the wind had blown all the leaves off the trees in the night, and was trying to blow the branches off, Pooh and Piglet were sitting in the Thoughtful Spot and wondering . . .

A. A. Milne

Nature Study, as a process, is seeing the things that one looks at. . . . The result is not directly the acquiring of science but the establishing of a living sympathy with everything that is.

"What Is Nature Study?"
(teachers' leaflet, New York State, 1897)

To my dad, who also loves to fish off the dock
and wonder . . .

Preface

FAWN ISLAND IS A RUGGED but poetic outcrop of granite in the heart of the North Woods. Draped with a shawl of juniper and jack pine, it rises out of the vast blue mistiness of Rainy Lake, overlooking a long sweep of open water sometimes lashed by wind and wave, sometimes still enough to reflect the starry depths of the universe.

From its rocky eminence I've watched shifting cloudscapes and read the weather of my own inward skies. I've paddled reflecting shores and drifted toward all the unnamed beauty of the world. In walking the island's trails, gathering its fruits, and splitting its wood, I've scratched under the surface of things, distinguishing more clearly the lynx of living from the lounge chair of existing. Each time I've come to the island I've discovered something new, have always left with a little more under my hat than when I arrived.

That battered hat, when off duty, hangs by the door of the Fawn's only dwelling, an old log cabin perched on the island's south shoulder. Inside, the logs are burnished a molasses brown from decades of oil lamps and wood smoke. Often those walls reverberate with late-night banter and repartee, the contentious noises of Scrabble and Parcheesi played to the death. During clan gatherings the days are alive with the hum of constant activity—dock fixing and berry picking and swimming, fishing expeditions for young swabs and old salts, the fixing of meals, the telling and retelling of family stories. For most of my life I dreamed of just such a place for sharing just such times.

But there are other times when I visit the Fawn softly and alone, times when I am feeling vaguely homesick for the earth and need time to wonder, to breathe the wind and drink the rain, to think what I think and feel what I feel, and sometimes, perhaps, to catch some of those thoughts and feelings in a net of words. Occasionally, too, I dip with others' nets—words that have inspired me or made me think or helped me to see the world a bit more clearly. Once in a while this seeing leads to a drawing, an image meant to capture a mood or moment or reflection. The capturing never quite works, but the effort is a pleasant one, and the memories are etched a little more deeply in the process.

Rusticating on an island roost can sometimes lead to Serious Thoughts. And sometimes not. Particularly if the rusticator is congenitally disinclined toward long spells of unbroken Seriousness. So, to quell any misperceptions and to evoke the proper spirit, I close with these sentiments from Isaac Walton, patron saint of fishermen and all lovers of watery places: "I wish the Reader . . . to take notice, that in writing . . . I have made myself a recreation of a recreation. And that it might prove so to him, and not read dull and tediously, I have in several places mixed . . . some harmless mirth: of which, if thou be a severe, sour-complexioned man, then I here disallow thee to be a competent Judge."

*If I cherish trees beyond all personal . . . need. . . it is because
of their natural correspondence with the greener, more
mysterious processes of mind—and because they also seem
to me the best, most revealing messengers to us from all
nature, the nearest its heart.*

John Fowles

THERE ARE CERTAIN PLACES in the North Woods
that seem to distill the essence of the whole country. A
chance combination of rock, tree, and water can be fo-
cused in such a way as to symbolize the entire character
of the land and illustrate the timeless forces that have
shaped it. Jack Pine Point is such a place. It was upon
discovering the point on an early exploration that we
first began to fall in love with the Fawn and decided
we must try to make it our own. On the tiny beach
cupped by the point, Kathy took a stick and scratched
in the sand, "The Woods," and I knew the long quest
was over. I also knew the point was, and always would
be, an enchanted part of the island.

The point is small, not visible at all from the open
lake east and south of the island; only from the quiet
back channel can the point be seen. Its glaciated, granite finger reaches out from the dark shoreline, extending west-northwest toward the summer sunsets. On its

north side is a shallow, rocky bay limned by cattails and bulrushes. Here the muttering black ducks hide their nest, and we leave them to their privacy.

On the point's south side lies the swimming beach. Protected from cold northwest winds and exposed to the sun for two-thirds of a summer's day, its waters warm up to what nearly passes for a comfortable temperature on Rainy Lake. Such a protected cove of sand is a treasure on any northern lake. Ours is a place not only for swimming and bathing, but for reading the traditional morning newspaper of the North Woods— evidence of the nighttime doings of neighbors heron, mink, otter, and the resident clan of beavers.

Marking the south end of the beach is Lifeguard Rock, a ten-foot-tall glacial erratic, emblematic of the many huge boulders scattered across the North. They were carried, often hundreds of miles, by mile-high glaciers, and dropped ten thousand years ago at their retreat. I can seldom look at this massive rock without recalling the day I saw Eric spread-eagled upon it, dozing in the sun, while a dozen turkey vultures in their undertakers' finest circled slowly above him. (We've not seen fit to let Eric forget it either.)

Meandering across the boulder's top is a narrow fissure, out of which grows a miniature birch tree. Were we able to watch for centuries on centuries, that tiny tree and its progeny to follow—aided by the ceaseless freeze and thaw and refreeze of winter—might someday split the great monolith in two. But for now there is a single Lifeguard Rock, whose name and brooding demeanor somewhat overstate the risk to life encountered by swimming at the tiny, sheltered beach.

The trail between sand beach and rocky point is

bordered by blue flag iris and meadowsweet. A footstep may find a sprig of sweet gale, and instantly the air is drenched with the incense of the North. I long ago discovered that a redolent twig or two placed under the strap of a Duluth pack can make the toughest portage trail a bit more pleasant. Or at least more fragrant. Another step toward the point and sweet grass is underfoot. One of the most sacred herbs of North American Indians, it has been gathered for as long as memory, carefully twisted and braided. Its perfume seems to last forever, no matter how old and dry the twist, and when touched to a flame it produces a smoke or "smudge" so sweet it is believed to purify any room, any gathering, any heart.

A few more steps and one is on the granite ledges of the point itself. Herbs are scarce now—a few corydalis plants flaunt their pink and yellow blossoms, a few sprigs of leatherleaf, a struggling red pine seedling. But there is one plant on the point that holds forth as boldly as a crest upon a shield, and that is the gnarled, drooping tree that, more than anything save the rock itself, gives the point its character.

Jack pines, with their relatively small stature and twisted grain, are not lumber trees, and were in fact long avoided by superstitious lumberjacks as bad luck. A jack pine won't win many beauty contests either. But to me this valiant old tree, solitary on its own rocky point, is as beautiful as a living thing can be. It symbolizes in its hunched and tortured silhouette all the fortitude and bravery to which anyone could aspire. In the calligraphy of its shape against the sky is written strength of character and perseverance, survival of wind, drought, cold, heat, disease; all made more challenging by the harsh restrictions and meager resources of its rocky hold upon the earth. Yet here it is rooted, here it stands. In its silence it speaks of a rugged humility, a wholeness and harmony, an integrity that comes from being what you are and belonging where you are.

For all these reasons and for others I cannot name, the little pine on its bare spit of rock epitomizes the wild spirit of the North to me, and the entire enterprise of life in a challenging universe.

If the beach is a place for sunny afternoons, then the point itself is the place for evenings. Tonight I sit against the pine's old trunk and watch the sun descend in a fiery spectacle through the drooping branches. The northwest sky fades from the color of crushed mulberries to a velvet indigo. Two loons cruise past the point and fling a hymn or two against the rocks. Water laps and gurgles softly in the crevices where the crayfish will be gathering as they move into the shallows. I add another log to the little fire in the stone ring and roll my bag out under the tree to watch the stars appear among its limbs.

Those limbs will not always be there. Someday the proud old bonsai will fall, may even be replaced by the little red pine to my side, and Jack Pine Point will become Red Pine Point. No matter, for what the point with its old tree *means* will never change. It will last as long as there are islands and stones and trees upon the earth; as long as there is someone to listen to water lapping against rocks, and to love the look of stars through wind-twisted branches.

Heaven is under our feet as well as over our heads.

Henry David Thoreau

Landscape is sacramental, to be read as text.

Seamus Heaney

A WARM, SUNNY, PERFECT MORNING. The local woodchucks lounge on their front porches in their BVDs, smoking cheap cigars, and scratching themselves. Squirrels seem equally indolent and content, pausing in their constant barber-pole spirals up and down the pines. Gulls float lazily off the reef, huge and surreal on the bright, reflecting surface of the lake. Far overhead a squadron of pelicans spirals slowly. Effortlessly they drift downlake on eight-foot, black-tipped wings, entirely in unison, appearing and disappearing as if by magic as they tilt in the sun. Finally they vanish completely into the wild and sprawling depths of Voyageurs National Park.

From just down the shore come the short, raspy quacks of mother merganser, the red-headed "sawbill" of the North, herding her rambunctious brood on their morning promenade among the islands. A spotted sandpiper skirts the water's edge, tail bobbing endlessly up

and down like a seesaw—the little "teeter-tail" of the shorelines. Out on the lake two loons begin yodeling wildly and are soon joined from the misty distance by other loons near other islands. More and more delirious the music becomes, virtuoso cadenzas flying back and forth in the still, concert-hall air. Rhapsody on blue. Then, as suddenly as the performance began, it ends, leaving only the soft, desultory calling of a vireo from the drowsy shade.

An egg-shaped gibbous moon hangs in the blue sky above pines and firs and cedars. Through their dark plumes spears of light occasionally pierce to the forest floor. Here the delicate and sweet-scented Canada mayflower blooms, wild strawberry and bunchberry, twinflower and nodding blue-bead lily. I spot a bright red wintergreen berry that has overwintered and pop it into my mouth, savoring the cool, wildwood taste. On the sunny ledges along the shore, purple vetch and yellow sweet pea are beginning to splash their colors across the rocks, and in the blueberry patch, bushes are just setting their berries. If the rains are timely we'll be busy there with our buckets come July and August.

I kneel down to catch the fragrance of a wild rose and a barred owl swoops by, trailing a cloud of crows in its wake. A noisy cloud. Crows seem never to tire of this ancient sport, deviling an owl from tree to tree, island to island. An old story has it that Gitche Manito, when the time came to clothe the first owl, had run out of feathers. He asked the other birds of the forest to contribute some of theirs and they generously did so, on the condition that when Owl grew his own feathers he would return the borrowed plumage. The feathers were never returned, to the resulting discomfiture of owls

7

ever since. (Crows were evidently appointed chief enforcement officers in cases of feather piracy, on the theory, one would suppose, that it takes a thief to catch one.) Now they move on—silent fugitive and raucous posse—heading across the channel, appropriately, to Crow Island.

Having started this far down the shoreline path, I continue my circumambulation of the island. Here the secretive creeping snowberry twines among tiny fronds of feather moss. There a young scrub oak struggles to get started, barely holding up two oversized leaves. Beside it, a patch of bearberry, the old kinnikinnick, or North Woods tobacco of the woodland tribes, hugs the ground in a green, leafy mat. I duck under a moose maple branch and am suddenly confronted by an irate ball of feathers, tail spread, neck ruffed out, mouth wide open, hissing, and apparently intent on devouring me, or at least my leg. A male ruffed grouse, defending its nest and attempting to distract me from it. It works. I am distracted. But undeterred. I climb back into my pants and move along.

Arriving at the beach I find fresh tracks of heron, sandpiper, mink, and muskrat trailing beside the soon-to-open blossoms of blue flag iris, the "fleur-de-lis" of the voyageurs. Silvery minnows flash in the shallows. Leopard frogs leap from sweet grass clumps at the water's edge. I cross the boardwalk through the tiny marsh, redolent with sweet, damp smells, and follow the trail as it climbs the high granite dome on the north end of the island. This rounded ridge of glaciated rock is an example of countless such outcrops that run—sometimes for miles—through the forests of the Canadian Shield, from Superior's North Shore to Hudson Bay and west to the Churchill River country of Saskatchewan. Like the others, ours supports only a few hardy pines and little foliage, but it is thickly matted with dry, crunchy cushions of caribou and pillow lichens, occasional red-capped British soldiers on a dry, weathered stump. Here grow also the wispy, nodding harebells and the shiny, whorled leaves of pipsissewa, sometimes called "prince's pine," which seems to love such sun-dappled spots. Here and there are patches of exposed rock littered with thin, loose wafers of granite—the typical exfoliating erosion of such an outcrop. I step carefully as I cross the ridge, placing foot before foot on the narrow trail to avoid disturbing its fragile rock garden. Footsteps are forever among the lichens. I pause to breathe deeply, holding in my lungs the summer perfume of warm rocks and sun-dried pine needles.

Around the Fawn I follow the winding trail, noticing that little has changed dramatically since my last excursion. This in itself is a fine thing to notice in a world of frenetic and seemingly permanent impermanence. But an island is not an escape from the world. John Donne said famously that "No man is an island," a statement that has occasionally rattled around in my head while here on the Fawn. I've turned it over and over and have concluded that—well, who am I to argue with John Donne. But I think he got it wrong. I think people are very much like islands—we're scattered about, separate and apart, significant and unique in our aloneness, each with our own shores, our own shapes and identities. Just like islands.

But here's the deal—no island is an island. They're all connected under the surface, all lapped by the same waters, caressed by the same breezes, lit by the same sun and moon. Fawn Island, with its circular form and circular trail, its ancient greenstones and granites rooted deep in the mantle of the earth, its pines aspiring for the sky, is in many ways a living mandala—a symbol—not of isolation from but of connection to—to the earth, the water, the sky. Here a morning walk is an encounter with rock-ribbed reality, a quotidian world of primal rhythms—of slow growth and decay, erosion and evolution, weather and season. It is a world in which change, slow and cyclical and seldom dramatic, can be appreciated and even understood, imbued with meaning rather than absurdity. A world dramatically in touch with all the height, breadth, and depth of existence. In such a place existence itself can become a medium of revelation, as of course it must if any particular revelation is ever to be possible.

So this morning on the island, as on countless mornings past, trees grow and flowers bloom, ducklings hatch and pelicans soar and rocks erode. And woodchucks lounge in the sun—lazy and serene—and scratch themselves.

OUTSIDE THE CABIN, a red squirrel is gnawing ferociously on Jesse The Body. Jesse The Body is a moose skull with a full rack who was found years ago, along with his entire skeletal remains, beside a North Woods gem called Jesse Lake. Via portage trail and canoe, Jesse (skull and rack only)—garlanded with an eclectic assortment of drying socks, underwear, and tennis shoes—made his stately way to a new home, where he now fulfills roles both ceremonial and practical.

Ceremonially, Jesse is a sentinel, a bleached reminder of the wild majesty of the North. He lends, I fancy, a certain old-time panache to the cabin. He is accorded occasional greetings as citizens enter and leave the premises, and even receives the odd "ooh" and "ahh" from new visitors. That's about it for ceremony.

Jesse's more important function is purely practical, for his wide but ever diminishing antlers are a prized source of calcium for the chipmunks, red squirrels, and other small neighbors who find such supplies difficult to come by. They gnaw away, and the tines of Jesse's rack become shorter and sharper with each passing month. When Jesse is leaned just so up against the cabin wall, the gnawing sound can reverberate quite nicely, like a primitive percussion instrument played by drunken musicians with no sense of rhythm.

Although this effect can eventually lose its charm, Jesse himself never does, and we shall miss him when he is gone. He won't *really* be gone, of course. He'll just be running all over a six-acre island in the form of very healthy squirrels with very strong bones.

THE CABIN IS JUST THAT. A cabin. It is not a summer home, a log home, a lake home, a cottage, a lodge, a chalet, or anything else of a more upscale cachet. It is a humble cabin and, as such, is heir and assign to those few rights, prerogatives, and idiosyncrasies peculiar to cabins everywhere.

For instance, it sags a bit—but most seventy-five year olds carry on their own private arguments with gravity. The cabin was built in 1925 by legendary local craftsman Emil Johnson, who also created Ernest Oberholtzer's fabulous buildings on nearby Mallard Island. Emil perched the Fawn's cabin on the most felicitous site on the entire six acres, a high granite shoulder with a long view downlake, a fine observation post for sun- and moonrises, and subject to every breeze or rumor of a breeze on warm summer days.

To welcome those breezes he designed sleeping porches on either end of the cabin. They have no glass

windows but are screened on all three outer walls. The screens are protected by wood shutters that open and close with a rope-and-pulley system. To fall asleep with all the shutters open, caressed by breezes and lullabied by loons and moaning pines is a rare pleasure, one that never grows old. It is, I imagine, the way sleeping was meant to be, before it got improved.

Between the sleeping porches, the middle of the cabin accommodates the small kitchen and the living area. Here Emil devised top-hinged windows that swing up and in and attach to wire hooks dangling from the pole-beam roof. Between the windows are small bookshelves, and against the south wall are a hand-hewn pine table and chairs. In one corner of the "living room" (which also serves as dining room, spare bedroom, and coat closet) rests the woodstove, a popular location during cold, rainy, or otherwise less than clement weather. Beside the stove is a sign that reads: "Smokers and chewers will please spit on each other and not on the stove or the floor." This seems to work.

Similar valueless adornments hang from various walls here and there—most of them either maps or pictures of fish. The fish are represented in a number of attitudes. There are fish in repose, fish swimming, fish jumping, fish about to eat other fish or chipmunks or perhaps low-flying geese. I have known most of these pictures since childhood, from when they hung in my grandad's basement. In one corner there are a few photos of my grandad and Uncle Wilbur and me—holding fish, of course—pictures taken forty years ago.

In another corner hangs my grandmother's old Chinese checkers board, a cardboard contour relief map of ragged hills and valleys and jagged fault lines held together by yellowed Scotch tape. The Chinese checkers marbles reside on the shelf above, in a red one-pound Folger's coffee can with a yellow flag on it that says "4 cents off." The marbles have been in that can all my life and, I've been told, for decades before that. Every time I take the can down and hear the familiar rattle of the marbles, I think of my grandmother, of long summer days, the two of us sitting on the floor together playing game after game. It makes me wonder at the strange permanence of provisional, temporary things— marbles dumped into a coffee can more than half a century ago, the sound of them, the feel of prying off the lid, sensations that became an intimate part of the lives of three generations. On the roofing boards above my head are smudged fingerprints left, I suppose, by Emil Johnson in 1925. The nails by the door where we hang our coats and hats—how long have they been there? I wonder how can we know, as we move through life, which gestures, actions, which of the tiny puffs of air that we call words, will long outlive us and the momentary impulse that produced them, becoming part of the landscape of someone's life?

I just know that I prefer the old, taped-up Chinese checkers board to any new one. That the marbles will stay in the Folger's can. And that a nail in a humble old cabin is a fine place to hang your hat. Better than most.

Woods are wide places waste and desolate where many trees grow in without fruit and also few having fruit. . . . In these woods are often wild beasts and fowls. Therein grow herbs, grass, lees, and pasture, and namely medicinal herbs in woods are found. In summer woods are beautied with boughs and branches. . . . In wood is place of deceit and of hunting . . . of hiding and of lurking . . . Also woods for thickness of trees are cold with shadow. And in heat of the sun weary wayfaring and traveling men have liking to have rest and to heal themself in the shadow.

Bartholomaeus Anglicus,
"De Proprietatibus Rerum,"
late fourteenth century

THERE ARE TWO THINGS to be discovered in any forest: You. And the forest.

Or, put another way, there are simply two forests to be explored. One is composed of light and shadow, soaring trunks and reaching limbs and unseen roots, small green shoots amidst mouldering decay, hidden forms, and mysterious forces.

The other one is outdoors.

After a spell, if you are lucky and persistent, there may arrive instances when the two forests become one, when you sense that it's all one great woods inside and out at the same time. At such moments you meet your own shadow on the shadowed trail, hear your own voice among whispering leaves. You never completely explore the whole great woods, of course. But you

learn—circling, probing, listening—and eventually, over a lifetime perhaps, reach a functional understanding that corresponds roughly with reality. An intuitive knowing. And you feel at home in the forest. The whole Forest.

WE HAVE AN OUTHOUSE called "the Church of Peace." It is called "the Church of Peace" because I hung a rough, hand-lettered sign on its door that says "The Church of Peace." I hung the sign there because my father told me that in the 1930s his Great-Aunt Rhoda had named the outhouse on the old family farm the Church of Peace, and I always thought that was the finest name for an outhouse I had ever heard. So when we came into possession of the island and its lovely old outhouse, we named it the Church of Peace, in honor of my father and Great-Great-Aunt Rhoda and other ancestors and the essential spirit of outhouses everywhere.

There is, of course, something about an outhouse that is very like a church. Once consecrated, a true outhouse (nonportable, no plastic, metal, or porcelain) is almost always a place of serenity and seclusion, cool and dark, where meditation is the natural response. Songbird choirs sing from green lofts, while organ music soughs

in the pines. An outhouse is often located in a picturesque setting and becomes an integral and beloved part
of the landscape in which it resides. Inspirational yet
functional, outhouses represent some of the most lyrical expressions of human architecture on the planet.

The Church of Peace is a two-story outhouse,
although that description implies an untidy physical
arrangement that does not, in fact, exist. It is simply an
unusually tall and noble structure built against a granite face that falls away to a low, leafy area. Where nor-
mally there would be a hole in the ground there is just
the lower story of the outhouse itself, extended out and
back with an auxiliary chamber. Far above are the steps
and old wooden railing, the cantilevered deck, the over-
hanging roof, the sanctuary. Cedars stand like deacons

beside the steps. Birches vault gracefully overhead.

Beside the door and its appellation is another small
sign that reads "Pay Toilet, 5¢." Under it is affixed a collection plate (catfood can) that has had fourteen cents
in it for several years. What Samaritan left this donation
we do not know (none of the regulars being the generous sort). But the treasury remains, saved as a mainte-
nance fund or perhaps toward missionary work. Other
gifts are also occasionally received, mostly consisting
of pinecones, leaves, twigs, and similar offerings left by
envoys from the small and furry denominations. I once
returned to the island in February to find a red squirrel
nest two feet in diameter, made of fur, feathers, leaves,
and cached with dozens of jack pine cones. As there was
still room—almost—to use the aperture, I left things as
they were for the winter, appreciating the enterprising
spirit of our guest. With the warm breezes of May,

however, and after Kathy had a chance to admire the scene, our little squatter was gently evicted.

I suspect it is impossible to spend much time in an old outhouse and not emerge a better person for it. The opportunity for personal reflection, to humbly engage the existential realities of life in a rustic, cloistral atmosphere, is balm to the soul. I've seen old photos of our outhouse in its construction stages fifty-some years ago. To sit where others who cherished the island for decades sat and ruminated, in a structure carefully maintained but weathered as the rocks, to know the same rustle of leaf and sigh of wind, to see the blue sparkle of the lake through green boughs is to feel a deep and comforting sense of perspective.

I've found myself in Manhattan offices, in fine restaurants and limousines, and caught my thoughts returning to our little island bower. And I've often sat inside the Church of Peace and been unable to imagine anywhere else I'd rather sit.

*The sun, with all the planets revolving around it and
dependent on it, can still ripen a bunch of grapes as if it
had nothing else in the universe to do.*

Galileo Galilei

FIRST LIGHT, and I sit on Moonlight Ledge—now
Sunrise Ledge—with a cup of Earl Grey tea. All is silent.
Nothing moves, no sound but the hushed voice of the
lake. The night was restless, roiled by winds and dark
dreams. Old doubts stood brooding watch, asking hard
questions: Will there be more books? Will things pan out?
What about college tuitions? Will we hold the island?

Now the day arrives in utter simplicity, out of the
bounty of the universe. No tormenting ironies, no trou-
bled winds of sky or spirit. There's a pale pink scarf of
fog wrapped loosely around Gull Island. As the sky light-
ens further, the star—our star—is just about to climb up
over Dryweed Island, having already risen over Namakan and Lac La Croix, Crooked Lake, Basswood, and
Knife, Saganaga and Gunflint, over Pigeon Falls and
Grand Portage and Superior, before that over Sault
Sainte Marie and Niagara and Manhattan and Fire
Island and Montauk Point.

Now the star trembles and pulses on the horizon, over Dryweed and the peninsula, red and fiery, and I can almost sense the turning of the earth from my little rock ledge, my own vantage point on the cosmos. I can imagine the familiar pink shorelines of Dashwa aflame, the white granites of Pickerel and Kabetogama shimmering. The sunrise begins to drip honey on the pines of Half-Mile Island and Gull's fog scarf is now the color of buttermilk. A song sparrow decides to break the silence, and soon a grouse drums from back in the jack pines. The rip of raven wing overhead. The first mewling of gulls from the reef.

Beside my knee a wild rose stands spangled in dewdrops, each drop holding the sun within it. I think of the night's doubts—concerns and problems that do not flee with the daylight. Still . . . next to a rose sequined in tiny sunrises, prospects seem less daunting. I look down the lake, down the old Voyageurs' Highway of toil and sweat and dreams, past Sandpoint and Steamboat to the Brule. I take a sip of tea. And I hold the sunrise, too.

WHEN WE FIRST CAME to the island, we acquired, among other artifacts, the only dock on Rainy Lake in need of regular mowing. In the real estate literature it was termed a "floating dock," but this use of language was simply a lively jest, as it was quite obviously more a sinking dock than a floating one. Its submarining tendency kept the surface of the dock constantly moist, and this, along with decades of decaying pine needles, leaves, moss, dirt, and rotting wood, made the dock a fertile field for any number of crops. Raspberries and cinquefoil, multiple species of sedges and grasses, sweet gale and horehound and mint, and a couple of young trees all made a home of our dock. Frequent visitors included otters, herons, and mergansers, who left their own contributions to the biotic community.

With such a unique and picturesque dock, it seemed a shame to change anything. But with the customary concerns about rusty nails and missing boards and neighbors who might fall through and not be seen again, we began the slow process of rebuilding the dock, plank by plank.

But we kept the skeleton, the old timbers. Even though it would have been easier to replace the entire dock, it seemed important to keep the massive old white pine logs on which it was built. A century or more they had stood in these woods, and for at least half again that span they'd been moored to this shore. Built by some long-ago craftsman, I know not whom, they were a direct link to the past, to the history of the land, the island, and the people who had been here for decades before us. A dock built on these timbers had meaning and character far beyond what any modern structure set on milled 4 x 4s and eight-foot flotation blocks could have.

So we simply found some small, algae-encrusted chunks of Styrofoam that had washed ashore, cut them into manageable shapes, and submerged them beneath the dock, wedging them between the old waterlogged timbers. The effect was immediate if not breathtaking. The old dock rode higher on the water, and with each passing summer has risen higher still, as the great logs gradually dry and begin to float once more. Meanwhile, I grudgingly continued to replace the old planks, by ones and twos and sixes and nines, until today most of the visible dock is new. And trustworthy. And rather normal looking. And not a very welcoming place for raspberries.

But I sometimes lie down upon it and peer through the cracks at the ancient tree trunks that support it and listen to the lake chuckling among them as it has for so long. I imagine all that the dock has meant as a gathering place on the water—the laughter and storytelling and swimming and cool drinks and hot coffee. I think of all the storms it has known, all the gaudy sunsets and star-strewn nights. I think about the years to come and memories yet to make. And listening to the lake muttering softly among the old timbers, I smile.

My soul can find no staircase to heaven unless
it be through earth's loveliness.

Michelangelo

The earth, that is sufficient,
I do not want the constellations any nearer,
I know they are very well where they are,
I know they suffice for those who belong to them.

Walt Whitman,
"Song of the Open Road"

L ATE EVENING, and I paddle among the islands, gliding between Mallard and Crow, back home toward the Fawn. The western sky hoards its last gold like a miser. Whitethroats and veeries call, leaving the tracks of their songs across the hush of evening. A loon wails from north of the Fawn, another answers from the south, and they begin an antiphonal chorus, the rock shores reverberating wildly with their echoes. The canoe pushes its breast against stained waters, and I fly into the reflections of the sunset.

Gradually the colors fade. Sky and water darken to a deep indigo. A crescent moon crowns the scene, earthshine of the old moon cradled in the new moon's arms. Following the sun, it too slides away, and, as the night deepens, the wilderness of the stars is revealed, the brightest ones—Vega, Deneb, Altair—dancing on the water.

Fawn is now only a dark and looming shape against the stars. The air is chill and I shiver, pulling the collar of my wool jacket just a little higher. Still I remain, unwilling to let the evening go. If ever there was a time for exploring the shadow shores of awareness, for probing the soundless depths of thought, this is surely it.

But no thoughts come. They seem to have slipped away as irretrievably as the sun and the moon, and, as the loons suddenly desist, there is only the silence of the stars. That, and a profound sense of harmony—what the Navajo call "hozhro," the feeling of being in tune and at peace with yourself and all that is. This, of course, is enough. In fact, it is everything.

I dip the paddle and slide past the bulrushes, past Jack Pine Point and toward the old dock, toward the soft yellow glow in the window of the cabin and the warmth of the woodstove.

Hozhro. Harmony.

My hands are cold. When I reach the dock and pull the canoe up, the gunwale slips from my grip, and the boat thumps against the old timbers. The sound cannot be loud but seems to echo over all the lake. It seems to shake the stars.

A loon calls once. Twice. The stars regain their equilibrium. I climb the stone steps to the cabin, open the door, and breathe in the cabiny smell I love, and I know that for this evening, hozhro is undisturbed.

Dow all the good to my Nighbor and my Self that I
can and Dow as Little Harm as I can help and trust
in God's marcy for the Rest.

Daniel Boone

THERE MAY BE NO MORE fiendish alarm clock on this earth than the Crow Outside The Screen Window. Unless it's perhaps this morning's design variation, Multiple Crows With Babies. True, I don't much like buzzers or beepers or jangling bells. And I definitely don't like clock radios, whether they're set to Mancini or Mantovani or Brahms's Second Symphony or Hot Country or Classic Rock. Aaargh. And I *especially* don't like to awaken to *happy* or *angry* or *shocking* Radio People flapping their lips about . . . anything.

But crows are relentless. Merciless. They can't be turned off or even turned down. They have no snooze button. And they are smarter, more purposeful than Radio People. It's clear they could be cawing, cater-wauling, and bawling anywhere else in the woods, on any uninhabited island, for instance, but something about a cabin with an open window shutter is irre-sistible. It triggers a powerful response in crows that will not be quelled except by the actual ambulatory movements, curses, and coffee-making efforts of the inhabitants.

All right, ALL RIGHT! I'm up already.

In truth I'm rather fond of crows. I admire their bright eyes and brassy attitude, their curiosity and playfulness and devotion to their youngsters. I'm just not so fond of these qualities at 5:30 A.M. But after a short hike to the Church of Peace and a few draughts of North Woods air and the aromatic promise of perking coffee, I'm feeling better. I've just finished thoroughly burning the toast and am in the process of sharpening it, when there is a knock at the screen door. There I find Ted and his best friend Eric, coproprietors of Gull Island and our next-door neighbors. Ted is clad in mackinaw and plaid cap; Eric, a gentle creature who looks to be equal parts dog, wolf, and bear, is wearing his usual outfit.

"Eric, would you go lie down, please?" says Ted as they come in. Eric finds the rug and parks himself there. "Thank you," says Ted as he sits at the table in the streaming sunlight.

"How about a cup of coffee, Ted?"

"Well, we can only stay a minute or two—we're headed to Ranier. Maybe one cup, thank you."

I return with the coffee (and a biscuit for Eric) to find that Ted, as is his happy custom, has brought a gift—a loaf of the world's absolutely best homemade bread. He makes the dough with a secret recipe and process that includes honey and west wind and sunshine, and he bakes each loaf in a coffee can. This results in a perfectly cylindrical loaf of bread with sunlight and balsam breezes inside it and a mushroom cap at one end.

When Kathy is here we'll return the favor with some strawberry shortcake or a piece of blueberry pie, but we will never quite catch up.

Ted mentions being up at 3:30 A.M. baking this latest batch of bread.

"What the heck were you doing up at 3:30 A.M.?" I ask.

"Baking this bread," he answers.

Right.

Ted, in addition to being a master baker, is a former regional editor of *Time* magazine. In retirement from that position, he became the founder, publisher, editor-in-chief, star reporter, and columnist of the *Rainy Lake Chronicle*, a weekly newspaper published in the lakeside village of Ranier. He knows more tales about Rainy and its history and inhabitants than almost anyone, dating back to his earliest boyhood summers at the lake in the 1920s as a guest of Ernest Oberholtzer. "Ober," long-time president of the Quetico-Superior Foundation, was Sig Olson's early mentor and a legendary warrior for wilderness. He was also, for fifty-some years, the sole full-time inhabitant of neighboring Mallard Island, nestled between Hawk and Crow and the Gull. Ted's devotion to Ober and his memory is profound; his roots in these islands reach down to the middle of the earth. He is their spiritual overseer, their story-keeper.

There are, of course, plenty of stories about Ted himself. For instance, an occasional passerby might idly wonder at the somewhat incongruous sight of a railroad caboose parked on the bare granite of tiny Gull Island. The explanation is simple enough—it's Ted guest cottage. He brought it across the ice one winter. When informed by the authorities that it was unlawful to have a house or dwelling within one hundred feet of the water, Ted found the obvious and sensible solution. He got a couple of lengths of track, laid them on the bedrock, and put the caboose on them. Now it was no longer a house, but a parked vehicle. This evidently fulfilled the letter and spirit of the law, and no further complaints were heard.

Perhaps my favorite Ted story is a personal one. Bryan and I had decided to visit the island one January for a few days of feeding the woodstove and cross-country skiing. The night before, I called ahead to Bald Rock Landing just to check on the weather and ice conditions. As we approached the island the next day on our snow-shoes, pulling our pack sled, we noticed a large black beast bounding around the island in the three- and four-foot drifts. Arriving at the dock, we found Ted, snow shovel in hand, ruddy faced, sweating, and smiling. He had just finished clearing our path from the lake all the way up the steps to the cabin. With Eric's help, of course.

But this morning there's little time for stories or reminiscing. True to his word, Ted finishes his cup of coffee, and he and Eric head down to the dock. There, Ted lights his pipe and they clamber aboard his lovely old wooden tug-boat with the one-cylinder Volvo engine. *The Teahouse of the August Moon*, the boat is called. Down the lake they go . . . chuff-chuff-chuff-chuff-chuff . . . at a stately and somehow entirely perfect speed. A few hundred yards out, Ted turns from the tiller and gives a wave.

I wave back, then climb the steps up to the cabin where I'll throw out the sharpened toast for the red squirrels. And the crows. And I'll sit down in the stream-ing sunlight with another cup of coffee and reflect for a spell about neighborliness. And independence. About loyalty and love—for a small piece of the earth, for a lake, for an old mentor. And I'll have a slice or two—or four—of carefully toasted, sunlight-sweetened, wind-freshened Ted bread.

And the crows get none of this.

Is not the sky a father and the earth a mother, and are not all living things with feet or wings or roots their children?

Black Elk

But ask now the beasts, and they shall teach thee; and the fowls of the air, and they shall tell thee; Or speak to the earth, and it shall teach thee; And the fishes of the sea shall declare unto thee.

Job 12:7–8

OF ALL THE WINGS OF THE FOREST, whose voices and forms and movements enhance the beauty of these woods, my favorite is the chickadee. This morning they cheerfully flock around me as I pause in my woodsman's chores, tired and sweaty, my back against an old spruce. The air is full of the quick, fitful flutter of their wings and the sweet rasp of their voices. Tsic-a-dee-e-e-e? they inquire. "Gijiga-aneshii" was the name given this little bird by the Ojibwe. In the Eastern Woodlands he was "Ch'gee gee-lokh-sis." Little friend chickadee.

And a friend he is to anyone who spends time in the woods, always ready with a sunny greeting, bright eyed, fearless and trusting and innocent. Just being around chickadees is enough to make a grumpy man whistle. Or smile. Or see the world with a clearer eye. "Tsic-a-dee. Hello. How are you? Is everything all right? I wish I could help. Maybe I can. Tsic-a-dee-e-e-e?" With that sweet, insinuating question mark at the end.

It has been traditionally said among the ancient peoples of the North Woods and prairies that the chickadee, though very small, is very wise. This brave and cheerful little creature understands many of the mysteries at the heart of nature, and in the hearts of humans. There is a very old story that tells of these things.

Long ago, in the Beginning Times, the People were very young, just learning how to live upon the earth. But they learned well, learned from their relatives the animals and plants, and found that the earth was good, full of many Teachers and wise Helpers who encouraged them to live in balance and in beauty. This was also a time when the Evil Powers struggled fiercely to overcome the good. The Evil Ones were not pleased to see the People living happily upon the earth. They decided they would try to discourage the People, to make them lose heart, perhaps even destroy them.

And so the Evil Powers sent fierce, cold winds and snows and storms to break the People's spirit. And heat and drought. And times of famine. They sent floods to wash the People away. And fires that sometimes swept through the forests and over the prairies. The Evil Powers set all these things upon the earth to create suffering and sorrow for the People.

Finally they said, "Surely by now we have broken the spirit of the Human Beings. It is time to send a messenger to learn how things are with them and to bring word back to us." And so it was they called upon Chickadee, called him to their lodge and sent him on his mission to the People. When Chickadee arrived at the dwellings of the People he was invited to enter. He was treated with respect and given a place of honor by the fireside to warm himself. He was given food and drink and

anointed with a smudge of fat, the sign of plenty. He was marked with a dab of red paint, symbol of the mystery of life.

After these ceremonies and marks of respect, the People waited and listened patiently to learn why their little guest had come. When Chickadee had explained his mission, his hosts held counsel to formulate their reply. And this is what they said:

"Go back to the Evil Powers, Little Brother, and tell them this: that the Human Beings are still alive and hopeful and ever will be; that the People will not be broken by discouragement, nor defeated by storms and stress, nor vanquished by hunger, nor destroyed by hardship. Tell them that Human Beings will always live upon the earth, that we will remember the goodness of life, and that the world is filled with beauty."

This is the message that Chickadee brought back to the Evil Powers. And it is the message that chickadees have proclaimed ever since, from firs and pines and oaks and cottonwoods, in winter and summer, in blizzard and drought. And it is why, even today, to hear a chickadee is to feel a little braver, a little stronger, and to feel a smile upon your face.

And it is no doubt why, as I get up and stretch my sore back and return to my chores, I notice the chickadees aren't the only ones whistling.

IN 1908 A YOUNG HARVARD MAN named Ernest Oberholtzer was told by his doctors that he had a bad heart and less than a year to live. He did die young—but it was in 1977. In between those dates, "Ober" carved out one of the North Country's most fascinating and significant lives, spending most of it on a narrow splinter of granite in Rainy Lake called Mallard Island.

After that deadly diagnosis Ober headed for the North Woods, hoping that the clean forest air and sparkling waters might be a tonic. Or, if that were not the case, at least to see and experience the Great North Woods, the "last frontier," before he died. The tonic took. Ober grew stronger and stronger, paddling and portaging some five hundred miles throughout the North every summer until he was nearly eighty. In 1912 he and his Ojibwe friend Billy Magee made an epic journey to Nueltin Lake and Hudson Bay, through unmapped country unseen by any white man since 1772. On his expeditions, Ober armed himself with a camera and a violin, to photograph moose, caribou, Indians, and Eskimos and "to serenade the geese and the bear and the rising sun."

Over the years the Mallard became not only the staging area for wilderness canoe trips, but also for a lifelong campaign to preserve and protect the wilderness. He battled the powerful borderland industrialist Edward Backus to save the great Rainy Lake watershed from inundation by dams, became founder and chairman of the Quetico-Superior Council, lobbied extensively in Washington, and was vital in the passage of legislation from the Shipstead-Nolan Act to the creation of the Boundary Waters Canoe Area Wilderness, Voyageurs National Park, and Canada's Quetico Provincial Park.

All the while, the Mallard was home. Ober's Harvard training as a landscape architect under Frederick Law Olmsted, combined with his own whimsical spirit, resulted in an ever-evolving collection of fairyland houses on the island. Replete with trap doors, secret compartments, stone and wooden footbridges, cantilevered decks and porches, the buildings, still intact today, seem to rise fully formed from the pines and granite themselves. Along with two former houseboats—one an old logging wannigan and the other a house of . . . of . . . well, a different kind of house—these buildings were the gathering places for Ober's vast assemblage of friends and acquaintances from all walks of life. From Harvard, New York, Washington, Minneapolis, Chicago, and the North Woods, they came to the Mallard. The island was Ober's ally, his secret weapon in shaping the opinions of opinion shapers and legislators who would innocently accept invitations to the Mallard and leave as conservationists. The island and its buildings also held Ober's lifetime collection of books—some fifteen thousand of them—his photographs and maps, his violins and pianos and recordings, his taped conversations with Indian elders. (Ober's Ojibwe name, Atisokan, means story-keeper.)

In short, Mallard Island became a spontaneous eruption not only of rocks and trees, but of one man's mind and spirit. It was an extraordinary place of hospitality and dignity and fun and reverence and beauty. Ober called it his little "University of the Wilderness." Others called it simply a "magic island."

And it is much the same today. Ober is gone, but reminders are everywhere, his influence permeates the island. People still come to visit, still by invitation. They gather to think and to write and to play music and to listen. They come to find inspiration, to find peace and silence, even to find themselves. And Ober, gone these many years, somehow greets them all.

Perhaps the entire work of civilization is to take
the simple and make it complicated.

Thor Heyerdahl

To attain knowledge, add things every day.
To attain wisdom, remove things every day.

Lao-tzu

Where is the knowledge that is lost in information?
Where is the wisdom that is lost in knowledge?

T. S. Eliot

I'M IN THE KITCHEN fixing breakfast, boiling water for Earl Grey tea and instant oatmeal. In the other room the tiny radio is on, and someone is reading the Morning News. A famous person that I admire has died. A plane has crashed, and many not-famous people whom I might have admired have died. War continues in a dark and bloody corner of the world, and many are dying. There is a big scandal in the government, and honor is dead. Something we all eat is bad for us and will kill us.

The News is not good.

The News is never good.

I step outside to attempt to balance the News with the Olds. Our star has come up and is still on fire. There is still morning mist on the lakes of the world and the sun is still burning it away. This morning's sunrise means the earth is evidently still spinning and probably remains in its orbit, which means we'll continue to have warm, life-supporting days and starry nights, and the year will continue to be about 365 days long, with four seasons. There is still gravity. I'm on the ground, the lake lies like a flawless mirror within its shores. Birds have evidently remembered how to fly and how to sing and are doing impossibly beautiful things in the air and with their voices. It appears that photosynthesis continues in the leaves of dogwood and bunchberry and fly honeysuckle and wintergreen and saskatoon, jack pine and ground juniper and birch and cedar. This means that other forms of life will have food and oxygen and will continue as well. Including the red squirrel, little *Sciurus hudsonicus*, who is barking and squeaking and trilling at me on a limb overhead, his forepaws jerking into the air with the force of every squeak. The squirrel is still here. The island is still here. I am still here.

There is, of course, News on the island as well. I could go read the "morning newspaper" at the swimming beach. But even without that expedition there is the shifting of breezes, the rising and falling of winds, the changing of weather; there is the arrival and departure and passage of each bird and swimming creature, noticed or unnoticed by me. All over the island things are dying. All over the island things are living. Just like in the Big World.

Every morning, every day, is a mixture of the News and the Olds. And somewhere in the balance we stand. And breathe. And look and listen. And eat our oatmeal.

THE CANOE IS PERFECTLY DESIGNED for two seemingly contrary functions—floating upon the surface of things and getting beneath the surface of things. There is no craft more at home in the world of reflections—rocks, trees, birds, clouds, stars—or better suited for exploring the hidden shorelines of reflections within reflections. What does it *mean* that there are stars, more than the mind can comprehend? That there is a universe at all? What is the significance of a tree, a rock, a thing—*any* thing? What supports it? Who and what and where and why am I? What supports *me*?

Given time and the proper surroundings, a canoe encourages such questions. In fact, it gently insists upon them.

And answers? Well, you dip the paddle, you follow the bow, you become a part of the reflections. . . . And they reflect in you. And maybe you call them answers.

B ILL HOLM IS VISITING THE MALLARD, and the islands quake and the trees tremble. The estimable Mr. Holm is a moving mountain of personality—a blond-bearded, blue-eyed, Camel-smoking, scotch-drinking, story-spinning, good-humored Icelandic Viking from Minneota, who sees the world with the eye of God and speaks with the voice of Odin. I've been invited for dinner. Boredom will not be served.

I paddle over in the Princess, through the high, pure melodies of dozens of whitethroats, the little "killoo-leets" or sweet-voices of the Malecites. "Hear, sweet; Killooleet, Killooleet, Killooleet," they sing over and over again, the clear, sweet notes falling like beads of maple syrup upon the still evening air. A vireo calls again and again from the high birches. It is the smiling story of the Ojibwe that the Great Manitou long ago set this little bird the task of counting all the leaves in the forest, and this evening the Leaf Counter is faithfully at his task. "Be-zhik, neesh, ne-sui, ne-win . . ." with bare-ly a pause for breath. I round the north point of the Fawn and a song sparrow carols and trills brilliantly from the bushes, "Sweet, sweet, sweet, unintelligible-ible-ible-ible . . ." Off the bow a snapping turtle the size of a garbage can lid dives down, down, down, into the amber depths of the channel. Rainy Lake is alive in all its early summer splendor.

I arrive at the Mallard to find Bill parked in an Adirondack chair on the wooden deck, a Camel in one hand and a tumbler of scotch in the other. He beams as beneficently as the sun and rises in greeting. With Bill are Jean Replinger and Joe Paddock. Jean is the official program director for the Oberholtzer Foundation, but in truth she is a warm-hearted caretaker, den mother, and favorite aunt of all who come to visit the Mallard when she is there. Joe, soft-spoken and self-effacing, with a perpetual twinkle in his eye, is a fine poet and writer who's on the island to continue his research for a biography on Ernest Oberholtzer.

I accept a cool drink and an empty chair and imme-diately become aware of magnificent aromas emanating from the open kitchen-boat window behind us. Bill is cooking supper. He soon rises to tend to the prepara-tions. Despite the beauty of the evening and the pleas-ant conversation, I find myself increasingly bedeviled by the scents wafting from the window. It's possible this has something to do with eating my own cooking for the past fortnight, though I too am a fine cook as long as two conditions pertain: One is that I am cooking over a campfire, alternately sputtering, smoking, and blazing. The other is that all invited diners have been paddling and portaging, preferably in a cold rain, for ten to twelve hours. This seems to bring out the best in my cooking or at least in the appreciation of it. It is also helpful if the food, prior to preparation, is labeled Dinty Moore, Kraft, or Jif.

Bill Holm's culinary art seems unencumbered by any such limitations, and we are soon invited inside to feast on sweet corn, new potatoes, Italian bread, and pork cutlets marinated in a sauce of vermouth, cornstarch, chicken bouillon, and wild leeks. All prepared to perfection. On the side, a bottle of merlot. And Bill regales us with tales of Iceland, China, and Minneota.

Somewhere into a second bottle of wine, after the dishes are done, the kitchen boat cleaned, and a number of increasingly bad Ole and Lena jokes told, things have settled into a comfortable quiet. Bill mentions an interesting old book he has come upon, about an arctic explorer named Stefansson. Pausing for effect, Bill quotes Stefansson: "The arctic is a friendly place. If you think." The words hang in the air for awhile, suspended like cigarette smoke.

Later on, pulling on my jacket and about to leave, I pause for a moment. "What do you think, Bill, is the universe a friendly place?"

Bill winks and answers, "If you think." Then his eyes turn serious. "If you think."

Any man who pits his intelligence against that
of a fish and loses, had it coming.

John Steinbeck

BACK IN THE DAYS when I was guiding many wilderness canoe trips, I acquired a curious speech pattern that became labeled as "guide talk." I don't practice it much anymore, having fallen out of the habit. But it comes to mind now as I tend a bobber off the end of the dock, nursing a cup of coffee and a lack of ambition. The bobber, evidently feeling the same way, has flopped dead on its side. Upon finally checking the situation, I find I've set the rig too deep and snagged the bottom. For some reason this innocent snag reminds me of another, almost forgotten snag—Judy's Snag— and it is Judy's Snag that reminds me of guide talk.

Guide talk, in its essence, is the art of appearing to say something intelligent or authoritative without actually doing so. All good guides use guide talk, though of course not all use it well. It's a necessity, as important to have in your kit as dry matches or a pocketknife. Why? Because there are, on any trip, Curious, Anxious, or

Skeptical folk who want to know Exactly What Is Going On, when in fact you, the guide, don't know—or, more to the point, don't want them to know.

"How much longer is this portage going to be?"

Not much, the worst is over now.

"Do you think it's going to stop raining?"

Yes. (No date given.)

"Are there any No-See-Ums in the North Woods?"

I've never seen any.

Guide talk is almost always encouraging:

Don't worry, that food pack you're carrying isn't nearly as heavy as it feels.

This is just a little cloud-to-ground precipitation, nothing to be concerned about.

No matter how cold and wet you are, you're always warm and dry.

Lightning will almost never hit the same tent twice.

Bears don't usually eat people.

It is often instructive:

You can lead a horse to water, but you have to paint a barn.

A yellow bird never craps at midnight.

And sometimes it is simply used to pose an interesting question or conundrum to occupy restless minds during a long day of paddling. My favorite:

What's the difference between a duck?

"Wh-Wha . . . Huh?"

What's the difference between a duck?

"*Between* a duck?"

Yes.

"That isn't . . . that doesn't make any . . . "

This particular koan invariably divides all members of the human race into one of two categories—those who think it's the funniest thing they've ever heard (me) and those who think it's the dumbest (everybody else). It does, however, serve to break the tedium and give people something to chew on besides jerky.

I get up and check the line again. Maybe it's floated off the snag. Nope.

You see, the real art and secret of guide talk is in understanding the proper dosage and application. Which brings me to Judy and her long-ago snag. Judy was quite obviously a wonderful person. She was also very Curious, Anxious, and Skeptical. Therefore I stayed alert to the need for guide talk whenever she was nearby. Judy kept asking me to take her fishing.

"Could you show me how to catch a fish?" she asked, with a poignant combination of hope and uncertainty. I kept saying I would. I finally did. We bade the camp a cheery good-bye and set off to catch Judy a walleye or ten. We paddled along a shoreline strewn with boulders.

"Are those the Glacial Eroticas that you told us about?" asked Judy. I affirmed that they were indeed Glacial Eroticas.

Judy asked about walleyes. "Are they big? Do they have teeth?"

I replied that they grew to enormous sizes and had teeth like railroad spikes. We, however, would avoid such monsters. "The place I'm taking you just has nice, small walleyes, but lots of them," I said. "Still have their baby teeth."

We arrived at the fishing grounds and rigged up. "Let out a lot of line, Judy," I said. "It's deep here."

We began fishing.

After a minute or two, I glanced at Judy's end of the canoe and saw her jig dangling about three feet under the canoe. "You have to let out a lot of line, Judy, it's really deep," I counseled helpfully.

"Okay," she said, and her lure sank from sight.

"Oh! Oh! Oh! I-I've got one!" came the news from the other end of the canoe. I looked to see Judy's pole bent double, but it wasn't moving.

"I'm afraid you have a snag, Judy."

"A snag???"

"Yes, it's what happens when you let out too much line."

"But you said to let out a lot of line."

"A lot, yes, but not too much."

"How much is too much?"

"Too much," I said patiently, "is just a little more than a lot."

"But *before* you said I hadn't let out *enough* line," said Judy accusingly.

"Well, you had only let out a little. It's deep here."

"How much is a little?"

"A little," I explained, "is somewhat less than not enough, which is a tad shy of just right, which is right smack dab between not enough and too much."

I thought Judy was going to cry.

"It's okay," I said, "let me show you."

We got loose from the snag, by the expedient of breaking the line. Judy looked, if possible, even more miserable.

We tied on a new lure and began to fish again. The fish did not bite.

"I thought you said," commented Judy, "that there were lots of fish here."

"There are," I answered. "Zillions of them. You have to be patient when you're fishing."

"How long do you have to be patient?"

I thought better of answering that one.

We were patient for a long time.

"Why aren't we catching any fish?" Judy asked eventually.

"They aren't biting," I answered.

"Why not?"

"Well, I-I don't know."

"You don't *know*?"

"There are a lot of things we don't know about fish," I said, somewhat defensively.

"How smart are fish?" asked Judy, squinting at me intently.

I thought better of answering that one, too.

Eventually we headed back to camp. The stringer was empty and the guide did not smell of fish.

We reached camp and I stepped out onto what turned out to be a less than stable rock. A moment later I was in the lake. Judy helped me out. We pulled the canoe up. I was dripping. I was cold. "Think I'll change clothes," I muttered.

Judy cocked her head slightly, and I thought I saw a hint of a smile. "No matter how cold and wet you are, you're always warm and dry," she offered pleasantly.

A bit later, as we all cooked supper, I grabbed the frying pan without a hot pad.

Judy found a bandage and some salve. "You can lead

a horse to water," she said as she taped me up, "but you have to paint a barn. And, you know, a yellow bird . . ."

It is probably not necessary to add that Judy remains Skeptical to this day, particularly around me. (Though her curiosity and anxiousness seem to have eased up a bit.)

Now, about this snag . . . A good jerk, a "snap," and I re-rig, careful not to set out *too much* line. The light is fading, tree frogs beginning to trill across the channel, but I'm not about to quit just yet. I'm sure if I'm patient just a little longer . . .

Oh . . . and the difference between a duck?
    Well, as Judy knows, that's no mystery.
    One leg's the same.

A MAJOR PORTION OF MY TIME at Rainy—particularly when in the company of sons and nephews—is spent in search of fish. Another major portion—particularly when alone—is spent in search of insights, or what might be called "wisdom." It occurs to me that these two things—fish and wisdom—are a lot alike.

Both are wherever you find them—shallow weed beds, cereal boxes; deep channels, old books; quiet bays . . . quiet bays. Both are likely not where you might have been told you will find them. Sometimes you have to sneak up on them. Those who talk most catch least. Both fish and wisdom seem to strike whenever you least expect them. Or just nibble. You catch few of either if you're moving too fast. Sometimes old ways are best . . . a cane pole off the dock. Still, there's nothing wrong with new technologies—bells and whistles, depth finders, computers—as long as you don't lose track of the goal. The real goal.

Oh, and both of them—fish and wisdom—start to stink if taken out of context and lugged around out of the water too long. Use them or lose them. Catch and release is good.

I can't stretch the analogy any further, but all wisdom aside, there's only one good reason to keep a fish and that is to eat it. To do that you have to fillet, or clean, it. I've been cleaning fish for most of my life. I cleaned all the fish at the resort on Kabetogama when I was a dock-boy. I've cleaned fish for my family—parents, uncles, cousins, children, nephews, nieces—and for all the canoe trips I've ever guided. In my athletic prime and with a well-honed knife I could fillet, skin, and package wall-eyes in thirty seconds each. I could remove all the cursed Y-bones from a northern pike and have it done in little more time. But I'm not as fast as I used to be. Hard to say why. I think I spend more time saying thank-you.

Over the years people have occasionally asked me to show them how to clean fish. This is a bad idea and I tell them so. Once you learn how to properly clean fish you have crossed a significant threshold that the average person is perfectly content to leave uncrossed. The threshold is inscribed "I don't know how to clean fish" and is invariably subtitled "And I don't want to learn, either." The person who does not know how to clean fish will seldom be asked to do so (and is fully aware of this fact) for the simple reason that the rest of us don't want to gag on a fish bone. Master the art of fish cleaning, however, and you will be called upon. Forever.

The acquisition of such arcane, esoteric knowledge is perceived by the general populace as admittance into a sort of elite club, a club from which there is no exit.

Like the doctor, the nurse, the cop, or the small-engine repairman, you can expect to be summoned for the rest of your life, at any time of need.

Membership is not without its rewards, of course. As in the aforementioned occupations one is regarded with a certain amount of awe. And . . . actually, I can think of no other rewards.

Except these. The jig dropped at dawn into mysterious depths. The plug flipped silently into the sunset. The insistent tug on the line, the dip of the net. The knowledge that in a matter of minutes, with a sharp knife and a campfire, a part of the lake will become a part of the fisherman. That with this knowledge come satisfaction and appreciation and a sense of responsibility.

And so I clean fish. More slowly than I used to . . . And I remember to say thank-you and I think about things. And while disposing of the remains I see in the dive of the snapping turtle and the swoop of the gull or eagle small curves in a great arc. And there is nowhere the arc does not go. Sky, water, earth. Me.

Of immortality the soul, when well employed, is incurious.
It is so well that it is sure that it will be well.

Ralph Waldo Emerson

Which way does a flame go, when it has gone out?

Gautama Buddha

The true name of eternity is today.

Philo

THERE IS AN OLD OJIBWE LEGEND, poignant and haunting, about a place called the Island of Souls. In the old story, death came to a beautiful young maiden in the morning of her days. Her love, a brave young hunter, was ravaged by grief. Finally, shaking off his despair, he determined to follow and find his beloved, believing she had departed only as the birds fly away at the approach of winter and that with due diligence he could find her.

Preparing himself as a hunter for a long journey and directed by the old tradition of his fathers, he traveled south to find the Land of Souls, leaving behind him the great star. Daily and hourly the air grew warmer, the ice thinner, the trees taller. Beautiful and unknown birds sang from the greenery, the sun stood longer in the sky, at night the spirits of departed men no longer danced wild dances on the skirts of northern clouds.

As he noted these changes, his heart became joyful, for they were signs of his nearness to the land of delight and his beloved. At length the hunter came to a lodge where stood an ancient and venerable man, a white-hair bent nearly double with age. As the hunter began to tell who he was and why he had come, the old man stopped him, saying he knew upon what errand he was embarked. The soul of a lovely young maiden had passed by a short time before on her way to the Beautiful Island, he explained. Fatigued from her long journey, she had rested awhile in the lodge. She had told the story of her young hunter's love and believed that he might try to follow her to the Lake of Spirits.

The old man told the hunter that if he made speed, he might yet overtake the maiden on her journey, but to follow he must leave behind his body and his weapons. The old one promised to keep them for him should he return. Following these instructions, the hunter left his body and weapons behind and continued on to find the blissful island. He came to the brow of a hill sloping down to a lake that stretched as far as the eye could see. Its shores were lined with beautiful groves of trees, and many canoes glided upon its waters. Afar, in the center of the lake, lay the beautiful island set aside for the residence of the good.

Upon the shore, the hunter found a canoe made of shining white stone, waiting as if ready for him. Taking the paddle, he pushed off and made his way toward the island. Soon he came upon a canoe like his own and in it found his beloved. Side by side their canoes drifted over the water, which he now realized was the Water of Judgment, over which every soul must pass to reach the Beautiful Island or into which the wicked must sink. The two lovers became frightened, for the waves began to rise, and around them many canoes went down. But the Master of Life decreed that they should pass safely, and they reached the island and landed full of joy.

Here the air was food, it strengthened and nourished them. There were mild, soft breezes, cool and refreshing shades, sweet waters, greenery and verdure on all sides. There was no conflict, no hunger, no mourning. Gladly would the hunter have stayed forever with his beloved, but the Master of Life spoke to him in the voice of a breeze, saying he must go back. His time had not yet come, he had not yet performed his allotted tasks in the land of the living. "Return to your people," he was told, "but in time you shall rejoin her, the love of whom has brought you here."

This story, unique in its beauty, is universal in its longings. When I think of an Island of Souls I tend to think north rather than south. I imagine the organ tones of wind in the pines, the calling of loons, the sweet violin notes of whitethroats. I smell the perfume of pine needles underfoot. In my mind's eye I see my grandad there, in his battered old fedora and his khakis, and I feel the warmth of his smile. I hear my grandmother's laughter. My Uncle Wilbur is there, with a barbed insult that feels like an arm around the shoulder. So is my Aunt Mary, her voice as gentle as the breeze. Sigurd Olson is cruising the shore in his Peterborough Prospector, and Elizabeth waves from the bow. There is room there for everyone I have loved, all who are a part of my feeling for the North Woods, but whose absence sometimes tinges these woods with a bittersweet hue like the last lingering tones of sunset.

But is there an Island of Souls? A heaven? An afterlife? What about purgatory? Reincarnation? I don't know. I have opinions on the various possibilities, but I don't know. And neither does anyone else now living. We have not been given the gift of the hunter in the old story, to travel to the beyond and come back. Millennia of universal beliefs, an innate longing and intuition of the soul, a sense of the beauty and balance and completion of Nature, all hint at eternity. Still, of life after death we cannot be sure.

But immortality is another matter. Of immortality we can be sure.

First, there is memory, and memory is no small thing. Through memory we can keep loved ones alive in a space that nothing else can fill. More important, we are able to extend someone's life beyond even our memories by simply being true to what we most admired and loved about them. In living out and continuing their virtues and values, we keep these qualities alive in us and pass them on and on again to others yet unborn. This is the whole meaning of tradition, of legacy and heritage, the importance of elders and mentors. To be remembered is one secret of immortality, and each of us is responsible for the immortality of those we love.

There is also, of course, the sort of immortality that has to do with genes and bloodlines, but this alone without the conscious humanity of choice and will, the sacrifice of parenting and the warmth of memory, gives little meaning to the term. The cloning industry, if one day it comes, will answer no ageless longings.

But there is another sense in which immortality is achievable and even assured, available at every moment, and it is this: "I" am alive. I feel the sun on my cheek and the wind in my hair. I see harebells swaying in the restless air and waves marching down the lake. I think. I feel. But I am hardly the only "I" to experience such things. Each and every person feels just as authentic, just as much "I," as I myself. One day I will die and my individual consciousness will withdraw into mystery. But beyond every death, "I" goes on somewhere else, in every individual consciousness, each with his or her own life to explore and embrace, his or her own family to love, own stars to see, own flowers to smell. No, it is not me reincarnated, but it is an "I" every bit as real and whole and legitimate as this one that I'm now experiencing. Each life born into and out of the world is as much "I" as this one is.

Therefore, I have immortality—in all these myriad existences called "I." This sense of immortality is available to me at any moment of empathy, when I am aware that every single person is an "I" and that a statement like "Love thy neighbor as thyself" is not so much a moral admonition as a statement of cosmic fact, as true as gravity.

We know these things instinctively. People dive into rivers to save drowning children they do not know; we contribute to charities; we worry about the happiness and survival of future generations we'll never see. We do these things because in our deepest selves we know that the "I" that really matters lives in each of us and therefore lives on after death. The essential "I" is immortal, certain as sunrise, sure as wings to a bird.

These are things that can be known about immortality. Here and now. As for what cannot be known, other dimensions, a realm of spirits, the province of mystery . . . well, I may choose to hope for an Island of Souls. I may even dream of it, and hold it in my heart, and smile when I cruise a pine-clad shoreline, a shore my grandad and uncle and I fished together, long ago.

71

> *"You know something? This is the first time that*
> *I've really felt that we were going to die."*
> *"I've known it all along."*
>
> Ethel and Norman Thayer Jr.,
> *On Golden Pond*

WHEN STAYING AT THE ISLAND, we dock at Lauren and Charlene Erickson's Bald Rock Landing on our trips to and from the mainland. This morning, on my way to run some town errands, I met Lauren futzing around the dock and asked him how he was doing.

As usual, his reply was not suitable for polite company. Lauren is the sort of human porcupine who is always ready with a bristly response. He would have you believe that he gets up on the wrong side of the bed every morning and that every day is a bad day. He has a cut-out of a black bear guarding his driveway. He has a sign that reads "Beware of Lauren." Call Lauren cantankerous and he'd probably figure you were trying to butter him up.

It's all an act, of course. Lauren is, in fact, the soul of northern hospitality, although it is unclear how much of this social virtue is due to the better angels of his nature and how much to the beneficent influence of Charlene.

A knock at the kitchen door finds Charlene eternally prepared with a bottomless pot of coffee and a stove overflowing with cookies or cinnamon rolls or some other temptation. A ready smile, a warm greeting, and a twinkle in the eye are all part of her charm, along with an innate instinct to care and to help. The fact is Lauren cares just as deeply and has countless ways of showing it.

I've long since lost track of how many times he's helped me with a balky motor, lent me a boat, turned his shed upside down looking for a tool or a nut or a bolt, towed a car out of the mud, and a hundred other kindnesses, all without a second thought.

And once in a while, in a quiet moment, I've been lucky enough to listen to Lauren talk about the lake and his lifetime upon it, his feeling for the great sweep of the land and its history, his decades as a commercial fisherman in the old days, the big storms, the big boats, the way the fishing used to be, the old-timers he knew. I've been privileged to view his astonishing collection of spear and arrow points and Paleolithic artifacts, which he seems to find as easily as others gather blueberries or mushrooms. At such times, there's a brighter light in Lauren's eyes, a different timbre in his voice. Something else changes as well. After such a conversation, and a cup of Charlene's coffee and a cookie, I climb into the boat and head back to Fawn Island across a lake that's somehow even broader and deeper than it was before.

There is nothing—absolutely nothing—half so much worth doing as simply messing about in boats.

Ratty to Mole, *The Wind in the Willows*

Well, what would you do with a brain if you had one?

Dorothy to Scarecrow,
The Wizard of Oz

THE GREAT THING ABOUT LIVING on an island is that you *have* to have a boat. Or, to be prudent, at least two. One may, of course, have a boat without an island, but this creates a problem. For the boat.

The problem is that a boat without an island is seldom really a necessity.

For those of us who love waters, a boat is never a mere *vehicle*. It is an object of art and desire and affection. A boat without an island may have all of the grand qualities that those of us who love waters love in boats. It may have charm, beauty, and classic lines. It may ride the waves like a dream, and, of course, place you in the ever-magic context of wind, sky, and water. But the signal characteristic of a boat that doesn't live at an island, that is not a necessity, is that it is *fun*. And anything that is always—or even mostly—fun lacks a certain . . . dignity. A seriousness of purpose or decorum.

Yet tie almost any stick-raft or dinghy, rowboat or

runabout, sailboat, kayak, or canoe to an island shore and make it your lifeline to the world, and it is imbued by circumstance with a dignity and decorum—even nobility—that transcends mere appearance and to which its mainland sisters cannot aspire. This craft is now your only connection to the realm of gas stations, hardware stores, sliced bread, and skim milk, and its dignity is palpable.

This morning the old Crestliner has evidently gotten up on the wrong side of the dock, and its dignity and nobility are (temporarily) obscured by the fact that it won't start. The fishing boat, meanwhile, is over at Mallard.

The editor from Boston and the illustrator from Ireland, visitors these past few days, need to catch a plane in International Falls and make connections to large and distant cities. I, captain of the boat and proprietor of the island, supposed expert on all things woodsy and watery and chauffeur to the airport, scramble frantically from ignition to choke to gas line to motor to battery, finally changing the latter. Nothing. Not a squeak, not a groan. I've never seen a boat—or any motorized contraption—so completely, thoroughly, dead.

At long last a good Samaritan in a living boat comes by, and we make our dash to the small airport. The plane is already gone.

An hour and a half at the counter and on the phone results in alternate plans that necessitate extra time and extra money, but eventually the travelers are on their way. They depart with warm words and forgiving smiles for the chauffeur, who then makes his way back to the lake to deal with the dead boat. Damn boat. Swine boat.

Arriving at Bald Rock I mention to Lauren and a

friendly bystander the morning's difficulties with the Deadest Boat In The World.

"You didn't happen to check if it was in gear, did you?" offers the bystander. "Won't start if it's in gear."

I borrow a skiff from Lauren. I go out to the island and take the boat out of gear. I start the boat.

I feel like an idiot, but that feeling will eventually pass. I *am* an idiot, and that will probably cause me problems for a long time.

To remain whole, be twisted. To become straight,
let yourself be bent. To become full, be hollow.

Lao-tzu

These trees shall be my books.

William Shakespeare

THE BEST TIME TO SPLIT WOOD is January when it is thirty degrees below. A close second would be February when it is fifteen below. A very distant third is June when it is seventy-five above—not that it is particularly unpleasant to split wood then, to be sure. It is simply that wood is not quite so amenable to being split when the mercury is above zero. Or above seventy. Particularly jack pine.

Jack pine is a wonderful wood to burn, if it is the only wood you have. Other noble woods—oak and hickory, for instance—despite their fine reputations, make a very inferior fire if you don't have any of them. So, most of the time, we burn jack pine. Dry jack pine burns brightly and fiercely in a woodstove and produces a very satisfactory symphony of pops and crackles and hisses. It also burns very fast—meaning several episodes of arising to feed the fire on a cold night—and it produces significant amounts of creosote, excellent for clogging chimneys and stovepipes.

But we are grateful for the heat and do not grumble and gladly gather, cut, and split the ever-abundant supply of dead and downed jack pine on the island. Sometimes we supplement the wood with a small cache of tightly curled jack pine cones, just to watch them open on the hot surface of the woodstove. Without a fire, the cones may remain sealed for decades, waiting. Fawn Island's even-age stand is likely the result of fire, nearly a century ago. Now the trees stand tall and sway and whisper in the wind, and make a fine nursery for the white and red pines and birches and cedars that grow up beneath them. But while jack pines may grow fairly tall, they seldom grow straight. And it is this characteristic I am dealing with as I split wood on a warm June day.

It is, as I say, against jack pines' nature, would seem to be very anathema to them, to choose a simple and direct route toward the sun. Instead, they twist and turn and corkscrew their way up, creating trunks that are full of personality and idiosyncrasy and resin, but which offer little help to the wood splitter. In the deep, brittle cold of winter, a good blow from an ax can often pop a short log in two. But in warm weather jack pine does not so much split as wrestle, trying to twist the ax handle out of your hand and break your wrist. Add to the stubborn, sinuous grain a few rock-hard, pine-knot booby traps, and one has a formidable opponent.

But if it is against jack pines' nature to be split, it is against my nature not to split wood. And since I am seldom around enough in winter to lay in a full supply, we have these summer skirmishes. I swing the ax. The jack pine tries to knock it out of my hands. We wrestle. I think of the old woodsman's saying that he who splits his own wood is warmed twice, note that it is not particularly useful on a hot day, and take off my shirt.

But in the midst of battle and occasional curses, I still admire the stubborn tenacity of the jack pine. I think of how well it epitomizes the wild, unforgiving North and all the places I have known in its vast stronghold. I think of my favorite tree down on Jack Pine Point and of Tom Thompson's magnificent 1916 painting "The Jack Pine," which somehow captures all the tranquil solitude, bold beauty, and harshness of the bush. I smell the sweet, pungent resin, the piney fragrance of the wood. And I have no complaints about splitting jack pine. Or about sitting by its warmth on a cold night and seeing in its leaping lights all the protean images of the North.

The tree, the waterfall, the bear, each is an embodied Force,
and as such an object of reverence.

Ohiyesa (Charles Alexander Eastman)

Be what you is, cuz if you be what you ain't,
then you ain't what you is.

Tombstone inscription,
Boothill Cemetery, Tombstone, Arizona

THE ISLAND IS NOT WILDERNESS. But it is close to the wild, near enough that the forces of nature are always close at hand. Even the cabin itself, old and weathered, seems more a part of the elements than a shield from them. On the island one is never far from sun and star and wind, the aspiring growth of green things. The flutter of moth's wing. The fall of raindrop. Here there are few diversions. Each aspect of nature, authentic and important, fills the moment in which it is encountered. Once truly noticed, a trailing stem of goldthread or twinflower winds not only through a green and shady sanctuary of the forest, but a green and shady sanctuary of the mind as well. Like the goldthread, I notice my thoughts becoming increasingly embedded in nature, in its forms and processes. And gradually, surrounded by the embodied forces of nature, I come to the realization that I am an embodied force of nature as well.

The full realization comes as something of a shock.

To be "an embodied force of nature," as I understand it, as rightfully entitled as trees and stars and bears and boulders, is no small thing. It causes one to look at things in a profoundly changed sort of way. And it entails some very fundamental freedoms and responsibilities.

To be impelled and animated and illuminated by the same Power that spins the planets and burns the stars and blows the wind and sings the birds and grows the pines is to be left essentially without excuses.

Forces of nature are always completely themselves and nothing else. They do not give up. In this regard the Zen maxim applies perfectly—there is no "Try," there is only "Do."

And, perhaps most important and most different from "normal" human behavior, forces of nature—authentic flowerings of creation—do not get in their own way. Most of us human beings—including this one—get in our own way constantly. We stop ourselves through fear or doubt or procrastination or insecurity or indolence or, especially and encompassing all of the above, ego. We allow our small self, the ego, with all its wounds and bruises, greeds and desires, to get in the way of the Big Self, the Self that draws as surely as the sun on all the powers of the universe, the Self that wants nothing but to Grow and Do and Be if only its nemesis would get out of its way. The Self that is, like the indomitable chickadee above my shoulder or the tree in which he sings or the star toward which it grows, a force of nature.

When I heard the learn'd astronomer,
When the proofs, the figures, were ranged in columns
    before me,
When I was shown the charts and diagrams, to add,
    divide, and measure them,
When I sitting heard the astronomer where he lectured
    with much applause in the lecture-room,
How soon unaccountable I became tired and sick,
Till rising and gliding out I wander'd off by myself,
In the mystical moist night-air, and from time to time,
Look'd up in perfect silence at the stars.

Walt Whitman

MOONLIGHT LEDGE. A last look at the sky before bed. One more view down the long, soft glow of the lake, bound by dark, silhouetted shores. A few deep breaths of juniper-scented, balsam-scented air. Suddenly the sky is ripped by a shooting star, light green, long tail, northeast to southwest.

There are, according to a small boy I know, two discrete sorts of shooting stars at large in the night skies: meat-ee-ers and veg . . . ve-ge . . . ve-ge-tar-i-ans.

This is the same scientist who divides trees into the botanical classes of Deciduous and Carnivorous.

Well.

The classifications work for me. What they may lack in erudition they more than make up for with a profound and honest sense of drama, of potential story lines. They are, in fact, useful reminders of Einstein's dictum that "the universe is stranger than we imagine; in fact, it's stranger than we *can* imagine."

True, a universe inhabited by grazing asteroids and meat-eating jack pines is passing strange, but is it stranger than one held together by "string theory," populated by wormholes and black holes and quasars and dark light and antimatter? And W. C. Fields and the Three Stooges?

What kind of universe would it be, after all, that would be too strange to imagine? Why, it would be none other than the one we gaze at by night and awaken to each morning and think we know.

I turn and follow the trail from the ledge back to the cabin. I tread on early Precambrian bedrock nearly three billion years old and find my way by the light of stars that may have died eons before my planet was born. And along the way I duck under the predatory boughs of a rather hungry-looking spruce.

A monk asked a Zen Master, "What does
one think of while sitting?"
"One thinks of not-thinking," the Master replied.
"How does one think of not-thinking?" the monk asked.
"Without thinking," said the Master.

MORNING, and in softly rumpled bedclothes of fog, the
island still sleeps. Nothing stirs, no breath of wind, no
rustling aspen leaf, even the chickadees and red squirrels
are still. I am up early to work in the woods, but those
plans must wait. To split the silence with an ax would
be intolerable. To rend it with a chainsaw, well . . . The
tools stay where they are, ax jutting at its jaunty angle
from the splitting log, chainsaw resting on the porch
bench.

Instead I make my way to Boulder Point to join the
silence. Along the shore the lake rises and falls almost
imperceptibly, too little to even chuckle in the hollow
places of the rocks. Half-Mile Island is hidden, the south
shore of the lake is nonexistent. Even Deer Island is just
a vague, dark hint through the mist.

I sit down on the granite, the boulder against my
back. The lichens covering it are soft and pliable in the
dampness. There is otter scat on the rocks, flaky and

granular and pink, a memento from last night's crayfish feast. Beside it, growing out of a wandering crack, is a pale pink corydalis, one of my favorite North Woods plants. This point on which it grows is exposed to everything, swept by wind and sun, frozen in winter, baked in summer. Here the waves have the fetch of miles of open lake, all the way from Brule Narrows. Yet every summer here the corydalis blooms.

This morning the crash and spume of waves is hard to imagine. Even the occasional boat wake from points unknown is absent. No whine of distant outboards— the fog itself seems to swallow up all sound, leaving a silence as deep as the moon's. I seek stillness of mind to reflect the stillness of lake, a gentle dip into the soundless depths of being.

But, as always, the goal of stillness is elusive. To clear the consciousness, to shut off the stream of incessant chatter that distorts a clear view of the world is difficult, even with practice. To face oneself without the shield of noise, even the noise of one's own thoughts, can be disconcerting.

The modern mind, the turn-of-the-millenium American mind—my mind—is something like a burnt bear in a cave full of hornets. Or, as ancient Hindu tradition has it, the mind is like a crazed monkey cavorting in a cage. Actually, a drunken, crazed monkey. No, worse than that, it's a drunken, crazed monkey that has Saint Vitus' Dance. To be even more accurate, it's really a drunken, crazed monkey with Saint Vitus' Dance that has just been stung by a wasp!

Well, bear or monkey, hornets or wasps, I am willing to sit here in the silence and wait for . . . silence. And eventually it comes. The bear and monkey relax and sit still, the buzzing swarm of thoughts, plans, speculations, and ruminations subsides, and I find myself on this island at this moment, on a weather-ravaged point with a boulder against my back. And I feel the presence of Things As They Are. And I listen to the fog.

Gradually, the sun begins to burn the mist away. Half-Mile Island emerges like a secret made plain. I get up to split some wood.

In the world to come, each of us will be called to
account for all the good things God put on earth
which we refused to enjoy.

Talmud

THERE ARE A GREAT MANY good and thoughtful
people who believe, quite sincerely, that the main purpose
of a cabin is to keep the rain off of your head. Though
widely held, this is a shallow and almost certainly erro-
neous interpretation of the history and function of cabins. I am quite sure that, could we pierce deeply enough
the mists of the past, we would find someone named
Waurnchf! saying to his mate (name unknown), "This
cave isn't working out at all—I can't hear a thing. Let's
build a log cabin with a planked roof so we can listen
to the rain."

Thus was the log cabin invented, to be followed by
further design evolutions like the tent, the lean-to, and
the Empire State Building. The latter, of course, is a
complete perversion of the original purpose, as only
those on the top floor can hear the rain, and, because
of a poor choice of building materials, they can't hear
it either. Even the modern house, huge and multileveled

and overly stuffed and insulated, is poorly designed to fulfill its original and most important purpose, allowing its inhabitants to listen to the rain on the roof.

But a cabin . . . Especially an old cabin, richly aged and mellowed like a fine violin . . . A night rain on its roof is the sweetest music, the most soothing lullaby.

And so I lie in the dark sleeping porch, under a pile of wool blankets, shutters wide open, and listen. The air is chill and damp, the breeze occasionally pushing a fine mist through the screens. But I am here alone, need not take Kathy's good sense into account, and the shutters will stay open. I pull the blankets up closer around my chin. The gentle tapping of raindrops on the roof, at first soft and hesitant, gradually more urgent, is now a steady drumming. Rivulets of rainwater cascade from the eaves in miniature waterfalls, spattering into puddles on the bedrock, drenching the clumps of grass and bearberry and corydalis around the roofline. Now and then the music of the rain is punctuated by the not very insistent mutterings of thunder and the soughing of breezes in the pines. Each gust shakes a new shower of raindrops from the trees, the trunks creak and moan, an occasional branch falls.

For a long time I listen, imagining all that the rain means, all the good it is doing on the island. Tiny hair mosses are uncurling from their tight brown coils and in daylight will be a luxurious green. Brittle caribou lichens on the north ridge are soaking it up, becoming soft and pliable. Polypody ferns and twinflower and goldthread, sarsaparilla and saskatoon, bunchberry and birch, all replenish themselves as the rain runs off the bedrock, finding clefts and crevices where the tiniest probing roots await. It soaks the thin layer of humus,

perhaps six inches of accumulated needles and leaves and branches and bones and skins in the ten thousand years since the glaciers, a natural sponge that dries all too quickly in a North Woods summer. Under my blankets I breathe deeply the lush moistness of the air and all the earth-locked scents, sweet and strong, that the rain releases, listening all the while to the subtle, shifting rhythms on the roof.

After a time the lullaby works its magic and I hear nothing more. Deep in the night I half-awaken. The rain has stopped, leaving only the dripping of the roof and the trees. By morning even that will stop, and I'll awaken to a clean and fresh-scrubbed world, glistening and full of life. And it will be good to feel a part of it.

SPECKLED ALDER. The name of a humble, moisture-loving North Woods shrub that grows in profusion in the low, wet area behind the beach. I'm clipping back some of the branches that have begun to stray across our little boardwalk, poking at eyes and knocking off hats.

I clip a protruding branch and toss it back into the brush and remember the last time I thought about the alder, several months ago. I had been asked to speak at a large, ecumenical conference and, as usual, had saved some time for questions and answers following my session. "Can you tell us," came the question from the back, "the name of your particular denomination or faith group, and would you give us your thoughts on the 'secular humanists?'"

I responded that . . . well, I wasn't actually too much of a "namer." That in my experience as a wilderness guide the easy application of names or labels usually did little to enhance understanding. Instead it was often a substitute for really getting to know someone or something, even an excuse for not doing so. (Concerned looks.) Many times, I went on, I had taken a group into the back country and had been peppered with questions: What's the name of this plant? What's that flower? That tree? That bird? . . . And had been dismayed to find that, five minutes later, not a single name was remembered, the plant or tree now disregarded or forgotten with the assumption that "we already know about that" when, in fact, nothing had been learned.

So I began a new approach toward names. I didn't use any. Or at least not right away. Notice this shrub, I would say. See how it only grows where its feet are wet—along the shore, in bogs and damp areas. It's not a tree, doesn't grow very tall, but look how its bronze bark is similar to that of a young birch, covered with these horizontal specks or lenticels. Now if you happen to have a mosquito bite or itchy place you can peel back that speckled covering and apply the moist inner bark to the irritation, and it will ease the itch. Notice, too, the deciduous leaves with sharp, serrated edges, but look—the bush also has cones almost like those of a fir or a spruce. . . . Now, what would you call this plant?

In my Chicago audience, several hands went up. "How about the wet-footed, mosquito-itch-stopping, deciduous-coniferous bush?" Sounds good to me, I answered. And what should we call someone we don't know or haven't met yet? Presbyterian? Methodist? Muslim? Hindu? Secular humanist? Or should we maybe just get to know them first? A ripple of laughter, some knowing smiles, a spattering of applause, a few frowns. But mostly the sense that, even a busy week later, most of them would remember the bush, the image of it in their minds, and maybe look forward to meeting one in person.

I move along the old planks of the boardwalk, reach up and clip another branch of the wet-footed, mosquito-itch-stopping, deciduous-coniferous bush. A good name. Before tossing away the branch, I peel off a little bark and rub it on a mosquito bite that's been bothering me.

DECADES AGO, when the earlier residents of the Fawn came here, the island was truly remote. These days, the subdivision and development of nearby shores has meant a proportionate increase in the local boat traffic. Warm summer weekends can be particularly busy. Still, there are many quiet times—moments to fulfill my longings for the wild and for the long, unbroken silences I've known in the bush. There are silent sunrises or mornings wrapped in battings of fog. Lazy afternoons when the thrum of dragonfly wings and the rustle of birch leaves are the only sounds to be heard. Sunsets when the golden air throbs and pulses to the ancient music of spring peepers and chorus frogs. And, of course, starry skies reverberating with the brilliant midnight cadenzas of loons.

All these things are a part of the mystique of the North and of the far-flung, lonely places that have become a part of me—Mirond, Amisk, and Namew, the roaring rapids of the Churchill, the Sturgeon Weir, the Fond du Lac, the Bloodvein.

But there is a time more than almost any other when, for me, the island recaptures its wild heritage—feels nearly as remote and isolated as it once was. It's on a day like this one, with a big east wind howling up out of the Brule, lashing the lake with a fierce and incessant energy. Few if any boats brave the lake, and all sounds are swallowed up by the wind, by the anthem of waves upon the rocks. The billows heave themselves like great, living things against the island, their spray and spume dashing up the cliff face and wetting the trunks of trees. Once or twice I've even stood there myself in a great, roaring gale until, like the pines, I'm drenched—just to sense the primal power of the lake, to feel a part of it.

There are occasions when I'm no great admirer of high winds, particularly damp, east winds that make canoeing, boating, and fishing difficult—that keep restless children and bouncy dogs cooped up too long in too small a space. But when it is only Kathy and I or when I am here alone, there is a certain satisfaction in knowing that, for a short time at least, there is nowhere to go and nothing that must be done. Then it is good to feel "marooned" on a secluded island, an island close to its past, enveloped by wind and the anthem of the waves.

That's a—I say, that's a joke, son! A flag-waver! You're built too low to the ground, son, the fast ones keep going over your head! I keep pitchin' 'em and you keep missin' 'em! (Nice kid, but a little dumb.)

Foghorn Leghorn

SOME PEOPLE JUST DON'T GET A JOKE. Sometimes I'm one of them. The universe obviously has a sense of humor or it wouldn't have invented the spruce grouse or the eelpout. Or the human being. And the human being wouldn't have invented the rake, left it lying on the ground, stepped on it, and whacked himself in the head with it. Again. (The rake of course has many different names—hubris, stupidity, greed, enmity.)

While there are many accomplished humorists in the woods—the Canada jay and the red squirrel, the bear, the otter, the crow—none of them has the knack for pure slapstick, the self-important pratfall, of the human being.

But . . . we are not always the authors of our own jokes. Sometimes we are just there—the chicken hawk with the baseball bat. Life, we learn, is chock-full of jokes. Lavish with tricks. And it's a good thing, too. They are useful little reminders that all is not pure tragedy. After all, we come into the world alone and in pain, and we leave the same way. In between are hardship and sorrow and strikeouts, leavened with love and beauty and base hits. But it can be hard to know just how to take your cuts . . . Until you feel a wintry grin on your lips. Sense the first vague awareness that, above or below or within it all, Something Else Is Happening. Something serious, very serious. But not *completely* serious.

Armed with your awareness, you learn not to dig your spikes too deep into the batter's box, and you're sometimes alert enough to see the high, hard one coming. And duck. Or at least take it on the arm and not in the ear. And you can even smile at the fact that the Pitcher obviously doesn't have complete control. Unless he was throwing at you in the first place.

Perhaps, as many indigenous cultures have believed, there are tricksters and Trickster-Heroes abroad in the world—the Great Hare (later to become Brer Rabbit), the Mammaygwessy, Manabozho, Coyote, Iktome (Grandmother Spider)—who, in the process of helping with creation, have left wrinkles in the fabric of life, screwups both major and minor in nature. Wild pitches.

Or perhaps not.

In any case, the way life is set up, stepping on a rake and whacking yourself in the head is still funny if it's not fatal. The world is humorous as much as vexatious. And even the little chicken hawk can get a hit if he keeps on swinging away.

Like the bright hair uplifted from the head
Of some fierce Maenad, even from the dim verge
Of the horizon to the zenith's height,
The locks of the approaching storm . . .

Percy Bysshe Shelley

His rash fierce blaze of riot cannot last,
For violent fires soon burn out themselves;
Small showers last long, but sudden storms are short.

William Shakespeare

THUNDERSTORM. I've been feeling it coming all day, and now, through the open screen door, I hear the first chesty rumblings. They growl across the lake and echo against the rock shores of Jackfish and Hawk and Crow. Aside from the thunder, all is quiet; there are no bird songs. Glancing out the window I see nothing moving, not even a birch leaf—but that will not last long. I pull on my rain pants, grab my slicker, and head out, the screen door slamming behind me.

From the dock a great thunderhead is visible, whipped-cream white on top, long streamers of ice crystals blown flat. It drags its belly over the islands to the north, trailing a dark skirt of rain beneath it. But it will not touch Fawn Island. Our threat comes from the southwest, where a squall line hangs like a windrow of snow before a roiling, black wall of clouds. The rumbling continues, more forceful, more frequent, the approaching blackness flashing within its depths. The topmost leaves of a birch or two begin to flutter, then to rustle insistently. The first cat's paw darkens the water. A few minutes more and two- to three-foot waves are already racing across the lake and slamming against the south shore of the island, just around the point from the dock.

The treetops now sway and dance, trunks creaking, branches rattling against one another. I check the ropes on the boats, buttoning up the old Crestliner. I haul the canoe off the dock and turn it over on shore, stowing the paddles underneath. Other than that, there's not much to be done but wait and watch and listen.

A pair of gulls appears out of nowhere and races over the island in an eye blink, riding the wind with abandon, startling white against the blue-black of the sky.

Everyone else appears to be tucked in, in dense stands of cedars and balsams, in tree holes, in lees, under rocks. In cabins. But I stay on the dock. Something about a thunderstorm sparks a sort of fierce jubilation in me, a feeling of being close to forces as primal as the earth itself, beyond the manipulations and controls of man. Years ago a brutal wind took down dozens of the oldest pines on the island, and I am still cleaning up the damage, trying to restore the old trail, fighting the insect infestation that can follow such a blowdown. Yet while I respect and even fear the power of great storms and mourn the damage they cause—to beautiful old trees, to property, to human life—still I would not want to live in a world without them, a world in which we did not see our own human limits transgressed.

The storms of Rainy Lake seem to get bigger and meaner than storms I've known in other places, though I'm sure it's the pure sprawl of the lake itself, its sharp-fanged rocks, and the fact that a wind can rile it so, which creates this impression. Now the wind rises another click or two, and the waves, capped with foam, pile still higher. The south shore roars with the crash of them. I pull on my rain jacket and tie the hood tightly as the first fat raindrops begin to spatter the old timbers of the dock, but it's not really raining yet. Now the unexpected happens. The wind increases still further, to a level I would not have imagined. Within moments the waves around the point have become monstrous, piling in from a direction in which they have only a limited distance to build. They hurl themselves with an almost living fury against the rocks, their spume flying all the way up to the cabin, into the trees. The trees themselves, which had been swaying wildly, now simply bend double with the force of the continuous blast.

I consider for a moment heading for the shelter of the cabin, but quickly decide I'm safer on the dock, around the corner from the violence of the lake and out from under the trees. Kneeling with my back to the shrieking wind, I can barely hear the ripping and rending of trunks stressed beyond their limits. Suddenly, the rain arrives in earnest. It rains pitchforks and ax handles, it lashes the island in horizontal sheets, and I give up trying to see what's happening. I simply hold tight to the dock and wait.

I don't have to wait long. A few more minutes and the storm has passed. I climb the steps to the cabin, clambering over three fallen jack pines as I go. Ten feet from the cabin door, a huge pine has gone down, falling away from the cabin with the southwest wind. The roof is littered with a few small branches, but no trees. We've been lucky.

I move down the trail to Moonlight Ledge and stand for a long time, watching in fascination as the great storm sweeps across the lake to Sand Point and Swell Bay and Bear Pass and Redgut. Tens of thousands of feet, perhaps eight miles high, it towers, its back side snowy white in the now dazzling late afternoon sun. It has left its violent mark, a mark that will be evident for years. I will especially miss the old pine by the cabin door.

As I turn back down the path to get the chainsaw and the ax, a loon wails from down the shore. I cast one more look over my shoulder at the billowing mountain of white, and it is absolutely beautiful.

> *I stood here, I stood there,*
> *The clouds are speaking.*
> *I say, "You are the ruling power,*
> *I do not understand, I only know what I am told,*
> *You are the ruling power, you are now speaking.*
> *The power is yours, O heavens."*

Pawnee Song to the Sky

THE GREAT BRONKO NAGURSKI HAILED from this area; after his retirement from professional football and wrestling, he returned to run a service station in International Falls. Once my grandad took me to shake his hand and the end of my arm disappeared. Completely. Bronko Nagurski's hand and forearm looked bigger than most *people* I had met. I couldn't take my eyes off of them. I was quite relieved and somewhat surprised when my own hand reappeared, intact.

Bronko Nagurski also had a cabin on the north shore of Lake Kabetogama, and sometimes on a fishing trip to that part of the lake we would pass by it and someone would say with awe, "There's Bronko Nagurski's cabin." Awe seemed to be a common and appropriate sentiment among our clan in those days. Awe at Bronko Nagurski and his cabin. Awe that someone might actually have a cabin in such a place. Awe

at the lake itself and the Great North Woods that en-
folded it and the wolves, bears, and bobcats that
haunted it.

I could hear this awe in the voices of my parents
and grandparents and absorbed it. Fifty weeks out of
the year we would speak of Kabetogama as one would
whisper of a mythic place—Olympus or Elysium or
heaven itself. But I had never been to any of those
places and knew they couldn't be any improvement
on Kabetogama. Besides, just saying Kabetogama was
to say as much as possible. The mere musical sound of
it evoked meanings that seemed intensely personal and
magical. The awe of my elders became in me a bubbling sense of wonder at all things Kabetogama-ish and
North Woodsy.

Looking back, it seems clear that that early and last-
ing impression of wonder has directed the course of my
life and career, has pulled me back again and again to
this area and on repeated odysseys throughout the
North Country. It is the hunger for even brief flashes of
the old magic—smells, sights, sounds, feelings—that
motivates me still, and when they arrive—always un-
expectedly—the world seems nearly as new and fresh
as upon my first glimpse of Kabetogama, my first breath
of its North Woods perfume.

Today I made a brief trip into International Falls. I
passed by the site of Bronko Nagurski's old gas station
and visited the new Bronko Nagurski museum. And now,
as I nose the boat into the dock, it suddenly hits me—
the old feeling of awe and wonder. I am docking at my
own island, my own cabin, on Rainy Lake, larger northern sister to my beloved Kabetogama. Between Bronko

Nagurski's old cabin and mine lie a few miles of water and the wild Kabetogama Peninsula. It seems impossible to think that, like the great Bronko, I have a place in the woods.

Of course, I still have my own forearms.

TODAY IS AN IMPORTANT DAY—the day I try out the new boat. But not without some trepidation. Through twenty-plus years of watery travel in Minnesota, Ontario, Manitoba, and Saskatchewan, my craft of choice has been the canoe. I've come to feel as comfortable in one—particularly in the Pistachio Princess—as in a pair of old shoes. Since coming to the island I've also maintained a succession of ancient motorboats, a deference to the exigencies of living on an island in a lake big enough to hold fifteen hundred of them, with more shoreline than the state of California. But now I'm about to baptize an entirely new and strange and different sort of craft, a sea kayak.

It does not look much like a proper boat to me. Lying on the dock, it looks like half a canoe with a hole in the middle, a cross between a boat and a donut, a skinny red pencil someone has sharpened at both ends, a . . . a *kayak*. Maybe I'm not a kayak sort of person. Maybe

this was a bad idea. It doesn't have seats; it has a cockpit. A small one. No thwarts. No gunwales. No place
for a Duluth pack. It has bungee cords. It has a rudder
you work with *foot pedals*, for crying out loud. But the
books and articles I've read by the kayaking cognoscen-
ti, the counsel of Helpful Friends, and the nice young
man at the store have all convinced me that this is just
the seaworthy craft I need to accomplish my goal—to
paddle the length of Rainy Lake and loop back around
the Kabetogama Peninsula through Namakan and Lake
Kabetogama. Alone. At least, I *was* convinced.

I'm not crazy enough to embark on this voyage the
first time I get into the kayak. Not any more. I tried that
method once, long ago on my first canoe trip—a two-
week excursion into northwest Ontario and into the
meaning of the word "stupid." And "miserable." (I've
done a few dumber things in my life, but can't think
of any of them at the moment. Maybe not.)

So this time, with the accumulated wisdom of years
and scores of wilderness trips, a different approach.
I've read lots of stuff about kayaks. I've talked to people.
I've purchased a good boat and accessories. Yesterday
I went into Ranier to meet Diane Wessels, owner of
Northern Cross Kayaking, who outfits and guides kayak
tours on Rainy and into Voyageurs National Park, and
she did not fall down laughing when I explained my plan.
She was, rather, most helpful and encouraging, exam-
ining the maps with me, suggesting good campsites and
scenic overlooks, discussing the differences between
canoe and kayak: packing gear, getting in and out, paddling, the posture needed to avoid developing a sore
lower back. Altogether it was a very positive meeting,
both of us feeling that my experience in canoes and on

big waters would serve me well in this new venture. But now I'm looking at this thing . . . And it does not look like a proper boat to me.

Maybe it will look better in the water. I haven't had it in the water yet, having brought it out to the island aboard the fishing boat. I slide it in. It floats. It seems awfully tippy. Following Diane's instructions, I bridge the double-blade paddle from just behind the cockpit across to the dock. I sit down in front of it. I brace my arms behind me on the paddle shaft. With my weight on the paddle and the paddle holding the kayak in place, I ease my feet into the cockpit, then up to my knees, then gradually slide all the way in until I'm seated. It *is* awfully tippy.

Exceedingly careful not to lean one way or the other, I gingerly push off, as fluid and graceful as someone in a body cast. But the ease with which the boat moves! A few cautious strokes and I am almost effortlessly in motion. I seem to have to only *think* "go" and go happens. I dig harder and hear the water bubbling and gurgling beneath the hull. I soon find that the long double-blade paddle serves not only as a means of easy propulsion but also as a brace and outrigger, and side-to-side stability is not nearly the problem it first seemed. The hesitant, tippy sensation quickly abates. A few more minutes and I've lowered the rudder and almost unconsciously begun to manipulate it with my feet, helping to hold the bow on line. No compensation or ruddering with the paddle, no J or C strokes, no switching sides. Every movement of the paddle contributes efficiently to speed and power and forward motion.

I circle the island with increasing confidence, once, twice, a third time; the shoreline flying by with a speed and ease I could never match in a canoe. I'm delighted with the feeling of being at loon's-eye level rather than above the water. Shoreline features—rocks, niches and crevices, leatherleaf and sweet gale, the exposed roots of pines—seem closer, more intimate than ever before. I also begin to adjust to the sensation that you don't so much sit in a kayak as *wear* it, a sleek, floating extension of yourself.

Stopping near the dock, I flip the rudder up and spend a few more minutes just playing, as I've done so many times in the Princess—backward, forward, spinning and turning, sculling, leaning out on a brace, experimenting with action and reaction, getting comfortable with this almost-but-more-than-canoe.

Finally, I glide into the dock, ecstatic. I am a kayak person! I'm going to like this boat! I brace the paddle behind me and onto the dock and after a couple of false starts drag myself gracelessly out. One leg's asleep. Once it awakens, I pull the boat up and stand looking at it for a few moments. It looks a lot better now, beautiful in fact. Just the boat for an eighty-mile trip around a sixty-thousand-acre wilderness.

I head up to the cabin with a smile, and, as I climb the old stone steps, I feel I've found a new friend.

Rules for Camping in the Great North Woods

Don't camp under trees that, in a big wind, may fall on your head or your tent.

Don't camp under especially tall trees that may attract a lightning bolt. By the same logic, don't camp on high, exposed rock outcrops.

Don't camp in low places where the rain will turn your tent floor into a lake.

Don't camp in dense, boggy places where mosquitoes and flies will be thick.

Don't camp on the leeward side of islands or points where mosquitoes and flies will be thick.

Don't camp on the windward side of islands or points where the wind may turn your tent into a parachute and your butt into an icicle.

Don't camp on beaches or sandy places because if you camp in sand once, you camp in sand for the rest of the trip.

Don't camp on rocky, rooty, uneven, or lumpy ground.

Don't camp on anthills.

Don't camp in a blueberry patch with large, brown cow-pies scattered about. (Those aren't cowpies.)

As this leaves virtually no place to camp in the Great North Woods, go camping anyway.

E<small>VEN ON AN ISLAND</small> it occasionally arises—the need to get away, to pitch a tent under different pines, to see stars through different branches. After scores of wilderness canoe trips, the itch sometimes becomes too powerful to ignore. Maybe head for the North Arm, hopscotch the islands—go deep into Swell Bay or Redgut. Or move down the south shore and into the park—through the great historical funnel of Brule Narrows past Soldier Point, into the backwater beauty of Kempton Channel, and along the magnificent cliffs and soaring granite domes of Anderson Bay. There is so much wildness and beauty in this vast Aegean Sea of the North Woods— the interconnected five lakes of Rainy, Namakan, Sand Point, Crane, and Kabetogama—that its siren song is nearly irresistible.

And so I've embarked on my little trip—nothing extraordinary, but different from others in one important way. This time I'm traveling by kayak. Tonight I'm camped on a bright scimitar of sand on the lee side of a piney island. High over the tent the wind roars like a waterfall in the tops of the trees, as it has howled all day on the lake. In fact, for the past several days the wind has blown steadily, and I've had a crash course in handling a sea kayak in rough waters. All went well, the craft handling the waves better than many a large motorboat and far better than an open canoe. (Hence the name, sea kayak.) After two days of practice around home, at gradually increasing distances in gradually increasing waves, this morning I set off. With each stroke across the open channel the fetch of the waves became greater, but after the initial consternation of paddling in seas not only taller than the boat but taller than my head, I began to relax and trust the craft and my instincts in managing it. After reaching the back channel and relative shelter of Grindstone and huge Dryweed, things calmed down considerably. Aside from the wind, the day was mostly memorable for the wolf.

It appeared as if by magic, out of nothing. It moved as if on nothing, the legs dancing gracefully beneath it but seeming not to touch the ground, supporting no weight. The motions seemed elegant, unhurried, but the movement swift. A gray or timber wolf, it was not gray at all, but rather a dusky, yellow brown, similar to a pack I once saw far to the northwest in the Fond du Lac country. Thirty, forty yards along the shoreline, in and out of a stand of cedars, it floated, and then it was gone, fading like a ghost back into the forest. And, like a ghost, difficult to believe in. There was no time for a picture or even binoculars, no one to share the experience with, no pointing fingers or excited whispers. Just a moment—a private, personal moment, with one of the ultimate icons of the North American wilderness. It's the first wolf I've seen in Voyageurs National Park. It may be a long time before I see another. The chances of meeting one traipsing across the Fawn are slim indeed, although one winter morning I did find a single track winding down our channel, past the snow-covered dock, and out across the open lake. But here in the depths of the park, moving quietly, watchful . . . expectant . . . there is always a chance. That, in fact, may be a great part of the meaning of the park itself, in terms of a myriad of experiences. None of them are guaranteed. But the chance is.

What, though, does it mean to have seen a wolf? It depends on who sees it, and how. Wolves are controversial—hunted, protected, feared, loved, hated. I've known of protesters who picketed a wolf exhibit because it showed wolves in the cruel act of killing a deer. Deer hunters sometimes squawk as well, because it's "their" deer the wolves are killing. Some see a wolf and see death in a fur coat, a bushy-tailed cattle rustler, Little Red Riding Hood's nemesis, and the arch enemy of the Three Little Pigs and countless little sheep. Some look at the wolf and see a true threat—if not to children and Red Riding Hoods, then at least to family pets and livestock and a North Country farmer's close-to-the-bone budget. Others see a symbol of intelligence and adaptation and harmony of creature and landscape, a living testament to the defiant persistence of the wild and a reminder of realities beyond our grasp. I must admit I am in this camp. But I wonder, as I settle into my sleeping bag, if some misty morning in the future, a great-grandchild of mine might be paddling a still-wild shore and come upon a wolf in the cedars and see . . . simply a wolf in the cedars.

Late in the night I half-awaken and listen for the wind. It's still there, high in the treetops. Three days now. Surely it will stop by morning.

Wrong. Today I learned something. Namely, that few things focus the mind with quite such alacrity as coasting on five- and six-foot combers that have been flung down twelve miles of open lake, then funneled through a narrows serrated with jagged teeth of granite. A parachute that won't open, maybe. But I haven't tried that. Today I did, however, find myself riding like a wild bird down the dark wings of a rushing windstorm.

First I dashed from island to island, never quite in the teeth of the gale but near enough—I thought—to gauge its force. Then it was time to make a decision. To make the crossing in front of Lost Bay, past the rocky ramparts that followed it, then on through Brule Narrows, or to hole up for another day or more in the lee of an island. I would get only one chance at the choice. There would be no opportunity for a change of heart, a 180-degree turn in the full sweep of the waves for a return to safety. I pulled out the binoculars and watched the lake, the crash of the breakers on the far rocks. I checked my gear, scanned the sky, gauged my comfort level with the kayak. I thought things over carefully. And made the wrong choice.

I don't know if the great pitcher, Satchel Paige, ever paddled a kayak, but he had the right philosophy when he said, "Don't look back. Something might be gaining on you." Besides increasing the fear factor, looking back also upsets the balance involved in riding the surf. Balance is good; upsetting is bad. Rather than looking back, it is much better to simply listen for the oncoming hiss of another racing comber, to feel the swell rising under the boat. Paddle hard to ride the crest, brace when you need to. Listen. Feel. Paddle, brace, balance. Don't look back.

Slowly the fangs of rock and their crashing surf slipped past. I hurtled by one other boat, a 24-foot Boston Whaler with a 150-horse outboard, beating its way up channel against the waves. Huge white wings of spray arched out from the bow with every thwack of the hull. The crew stared at me with a mixture of awe and concern, as if they'd encountered a not-too-bright emissary from another planet. They waved. I nodded and smiled, sort of. But the hands definitely stayed on the paddle.

Finally I passed Soldier Point, the wave-magnifying chute of the Brule behind me. Though the wind still howled, the chop of Saginaw Bay was now manageable. I paddled on as long as I could and finally made camp with arms that felt like legs and legs that felt . . . really stiff.

Now the tent is up, supper cooked, and dishes cleaned amidst the final challenge of a passing rain squall. I sit by the campfire enjoying a small, end-of-the-day reward—a sunset-lit double rainbow arching from one shore to the other. Tomorrow will be a better day.

Yup. Much better. Sunshine. Soft breezes. The intimate, mirrored beauty of Kempton Channel. Then, on down the south shore, past frowning cliffs and sheltered beaches, the majestic heights of Anderson Bay, where I stopped to absorb the panorama and eat a sandwich. From the great granite domes I caught a feeling of space and perspective I've known in few other places in the North. In fact this whole stretch of rock and woods and water seemed as wild and lovely as anything I've encountered on far-flung expeditions.

Late in the afternoon, as I rounded Rabbit Island and entered the border narrows below Kettle Falls, it became impossible (well, very difficult) to imagine going on past historic Kettle Falls Hotel—a soft bed with cotton sheets, a hot meal cooked by a real cook. Funny, ten years ago such a luxurious layover would have seemed like a hopelessly decadent and wimpy thing to do, a violation of my jack pine–savage, wilderness-roughneck code of honor. Now it just seemed like . . . a really good idea.

So tonight I lounge in sinful comfort—in a marvelous old edifice constructed in 1913, an integral part of the century's history in this wild corner of the world. One-time haunt of bootleggers, commercial fishermen, lumberjacks, prospectors, and fancy ladies, the hotel is still reachable only by water. It looks and feels much as it always has, right down to the steeply rolling wood floor of the old bar, which makes you feel you've had too much to drink even before you have.

But tonight the main attraction for this visitor is a soft bed in a warm, dry room, wet clothes draped over chair backs, window open slightly, a light breeze billowing the curtain, carrying the muffled thunder from the waterfall. . . . For a last few moments I peruse a booklet about the old hotel, the history of this "jewel of the forest," and finally drift off to the poetry of a 1930s advertisement:

Here is the nation's playground—manifold lakes, blue and sparkling in radiant sunshine. Winding rivers and pleasant streams. Boating. Fishing. And here also is good health, invigorating air, light and dry, and laden with the odor of Balsam and Pine. The absence of noxious weeds makes this a playground for the hay fever sufferers. The cool, sweet-scented nights are restful to all.

I left the hotel this morning to the lush, nostalgic sounds of Glenn Miller's "Canadian Sunrise" playing in the lobby, and the languorous mood lasted throughout the day. I stopped at the old mine on Mica Island and several other pretty spots. A couple of sun showers grazed me with the hems of their skirts. I arrived mid-afternoon at Torry Fish Camp on Kubel Island and could not bring myself to go farther.

What a graceful, lovely place! In the afternoon sun, the best of the border country seemed to have been gathered up and spread out like a banquet. Clean, granite sidewalks, laced with bright igneous intrusions, winding among clumps of ground juniper and bonsai red pines. Delicate corydalis, mosses, and lichens all added to the impression of a perfect Zen garden. A fine, secluded beach. From deep in the woods, a hermit thrush trilled his magic flute tones.

And so, half-bewitched, I made an early camp to soak up the beauty. I hiked and explored. I set up the tent and gathered firewood. I wrote a bit. And I reflected for a spell on the strange power of a paddling trip to distill and clarify. Days that are too often fragmented and incoherent begin to come into focus, as under a lens. There's a beginning, middle, and end to a day, all connected by a path—a path of waters. Islands and headlands pass, with plenty of time for the mind to take them in. Wildlife and plant life are observed, weather noted. It's not fast, it's not exciting, but in its very simplicity it allows for the coherent experiencing of one's day. Of one's life.

For a good part of this century, Finnish pioneer Lydia Torry experienced the days here on this wild and lonely point on Namakan Lake. She lived here with her husband Emil, a commercial fisherman, from 1924 to 1954. After his death she continued to live here alone, year-round, until 1979.

As the last soft glow fades from the lake and I burrow into my bag, the words of Lydia Torry echo in my head—words that are preserved in a plaque on the bedrock. They are as firm and grounded as the granite itself: "I got used to being alone. It's part of my life here. There aren't any hardships. I like the water, trees, most anything. I even like the stumps."

Slowly they slip past—all the place-names that haunted my earliest dreams. To say them aloud is still like reciting a poem learned in childhood: Blind Ash Bay. Lost Bay. Wolf Island, Pine Island, and Martin. Daley Brook. Nashota Point. Sugarbush and Cutover. The Grassies. All day long I pass them, each one with its own definitive shore and skyline, even its own smell. Its own memories. Yet how strangely different it all seems from the vantage point of a kayak paddled alone, rather than a fishing boat full of family.

Here I lost a big one. Over there, too. My dad was marooned one stormy night with my cousin and brother at

the foot of this big island. Here's where Kathy caught a fish on her first trip to the North Woods. This is the point where I took Bryan camping for the very first time. And that's Wilbur's Bay, a name found on no Geological Survey maps but etched into family tradition as my uncle's favorite spot.

Kabetogama is a big lake that can get rough in a strong wind, but today is thankfully quiet, and the miles and islands glide past with surprising speed. I had thought to camp on the lake this last night, maybe on Sugarbush or Harris—but, somehow, almost without my intending it, the paddle keeps stroking, the islands fall away, and only one island beckons. The Fawn. If I keep moving, I'll soon reach Echo Bay, home of Alan and Miriam Burchell's Moosehorn Resort, and maybe I can bribe Alan into a lift around the back side of the peninsula and back to Rainy.

After a day of blue skies, the last mile becomes a race between me and a thunderhead looming over Echo Island. I do not win the race. Finally I ground the kayak on Alan's swimming beach, safe but soaked. Moments later I'm basking in the warmth and ambience of the Burchell's living room/resort office, a room I first entered at the age of seven, when Alan's parents owned the resort.

With laughing incredulity and concern they ask about the kayak that has just come paddling out of a thunderstorm, all the way from Rainy Lake! They ask about family. Miriam rustles up some hot coffee. We talk about the resort and fishing and comfortable things.

The storm is soon past, and Alan proves to be a delightfully easy bribe. An hour later I am back at Bald

Rock, waving a grateful good-bye to Alan, then gliding over sunset-stained waters toward my campsite for the night—a little rock in a big lake, part of a vast inter-connected waterway that stretches from Superior to Athabaska. My little part of it. A part called Fawn Island.

> *There was once a snail who was run over*
> *by a turtle and rushed to the hospital.*
> *"What happened, how were you hurt?" everyone*
> *asked as they fussed over him.*
> *"I . . . I don't know," he stammered. "It all happened so fast."*

> *The Now in which God created the first man and the*
> *Now in which the last man will disappear and the*
> *Now in which I am speaking—all are the same in God,*
> *and there is only one Now.*

> Meister Eckehart

STAND ON A GRANITE LEDGE in the last light of day, beneath gray wings of clouds undershot with crimson. Geologists say the rock on which I stand is approximately 2.7 billion years old. Astronomers, still arguing over the age of the universe—this current universe, at least—have narrowed it down to between 12 and 15 billion years of age. And the possibility is still open that this is just one version of an endlessly repeating universe, expanding and contracting like the breathing of a sleeping bear. In which case the universe might really be 1,000 billion? 10 billion billion? years old. (The latest research, however, indicates an open universe.)

Time, I know, is Nature's way of keeping everything from happening at once. But what else is it really? Particularly, what is a "long time"?

I was recently looking through a family photograph album on the occasion of Bryan's birthday. It's *not possible*, I thought, that this was so long ago. And, of course,

I was right. It wasn't long ago. The old clichéd feeling is true and seems clichéd only because it is true and so many have felt it. On the island we have another photo album—one of our island treasures—an old, water-stained book that came with the cabin, filled with black-and-white photos of people from six or seven decades ago, people who loved the island as we do. In the pictures, we see them building the outhouse, building the chairs we still sit in, gathered around the same table, laughing, playing cards, the women in long, ankle-length skirts and aprons, the men in high caulked boots and lumberjack shirts. I'm sure all are gone now. It was a long time ago. A lifetime ago. So much has happened, so much has changed.

But what if a lifetime were not a long time, which it certainly isn't, geologically or cosmologically speaking? What if our normal, daily view of time is as skewed as a mayfly's, as self-referenced as the poor, injured snail in the hospital? What if five hundred years were not a long time? A thousand years only yesterday? Two thousand—the time of Jesus—just the night before last? What if our ancestors were not really our grandfathers but our brothers? Not so much our great-grandmothers as our sisters? Then time—or our common perception of it—would have no relevance to truth and reality. Would be, rather, an impediment to it. A veil. What if, as the old photo album suggests, the swift passing of time is an illusion, and "It couldn't have been so long ago" is true. Not a nostalgic cliché, but a flash of insight?

"Nothing lasts for long," says our galloping world, over and over again, every day, in a million subtle and not-so-subtle ways. We are left, subconsciously, with a hunger, an ache for things that do last, that are connected to the past. Or else what is there to strive *for*, to build a life on and to measure it by, to connect the days to weeks to years to decades?

Rocks help. Lichen-covered boulders that have witnessed the passing of birch bark canoes and the endless rafts of the great logging drives, that have stood still for a black-and-white photo with a young lady in a long, flowered skirt. Tall pines help, too. And deep-water lakes. And old cabins and things people have made with their hands. And above lakes and rocks and cabins, constellations that dangle over dark and unchanging shores.

I sit down on the bedrock ledge and light my pipe. I watch the smoke rise. I take a sliver of the rock and stick it in my pocket. I don't move for . . . a "long time"? Not, at least, until the stars—the very nearest of which has taken 4.3 years to send me its light (at 6 trillion miles a year), the farthest yet detected, maybe 12 billion years—begin to appear through wisps of cloud. The smoke rises toward the stars. From Deer Island a barred owl issues his bold, rhythmic challenge. The lake washes softly at my feet. And in the eternal, undiscovered present, I enjoy the time.

If a plant cannot live according to its nature it dies,
and so a man.

Henry David Thoreau

Those who contemplate the beauty of the earth find reserves
of strength that will endure as long as life lasts.

Rachel Carson, *The Sense of Wonder*

THERE IS A DISTINCTION often made in literature and common language between the "mind" and the "heart." While the language may not be anatomically or medically correct, I find that in the woods, on the island, the meaning becomes clear. And the primary distinction is this: where the mind thinks, the heart knows. And herein the potential for inner conflict is sown.

For what the heart knows is not what "the world" (the cultural world) knows; it is, in fact, often in conflict with what the world knows. And if you are far from yourself, your mind thinking as the world thinks, you cannot know your own heart.

This is the life many of us lead to one extent or another—a life that looks all right from the outside but feels splintered and disconnected on the inside. It is why it is so difficult to "follow your heart." How can we follow what we don't know? In the absence of this knowledge, "buts" and "ifs" intrude upon any quest or endeavor, and we become lost in comparisons or fears or disappointments.

I have always felt closest to my heart in the woods—beside a campfire or shoreline or under the whispering of trees. I know that I am far from alone in this feeling. For millennia, perhaps for as long as there have been people, people have gone to the wilderness—the forest, the desert, the mountaintop—to find God, to find themselves. Today, being unsure of God and less sure of ourselves, we may go with a more modest agenda. We go, we say, to find "peace and quiet."

The irony is that wild places, the woods, for instance, are seldom truly quiet. And they are not peaceful. To look into the woods and see peace is to not look deeply enough. The lake is not peaceful either. To look deep into its waters and see peace is to not look deeply enough. Everywhere and always the struggle goes on. For light, for life, for primacy, for survival. The chickadees struggle. The red squirrels and pine martens struggle. The wolves and deer. The junipers and jack pines. The small herbs and flowers. The turtles. The fish.

And yet, to look into the woods and waters and see only struggle and suffering is to *not look deeply enough*. Because beneath and beyond the struggle there is . . . peace after all. Beauty. Wholeness. For wild and natural things the fact that life is difficult—a struggle—is manifestly irrelevant. There are no comparisons made, no ways out to be found, no improvements sought. And because the struggle is irrelevant, it is not really a struggle. It is something else. It is . . . what is.

To observe and contemplate the natural is to see Life. That is, to see suffering, pain, and struggle—and to understand by extension that these conditions are not aberrations or curses that are visited upon our own lives but are an integral part of the way things are, the way the world works.

Perhaps, in the context of the wild, in the heart of the woods, we do come closer to our own hearts—hearts that accept as good and necessary the reality of struggle, of difficulty and death, the reality that all of nature accepts and is a part of. And there we may find our sense of peace. And beauty. There we can distinguish between what the mind thinks and wants and fears—and what the heart knows. There we may say yes to our own lives, in all their joy and hardship.

ICOME TO JACK PINE POINT AT NIGHTFALL, dark shoreline reflected in barely undulating water. The tree frogs and spring peepers sing from the little bay across the channel. The singing is loud—steady, throbbing. Now the American toads begin with their long, breathless trills, and the music becomes even louder, more eerie, more insistent. That music is nearly as old as animal life itself on this planet, an echo of the dim past as well as a living symphony of the present. It fills the air and it fills the night, and it seems to reverberate back from the sparkling dome of the stars themselves.

Under the moonlit water, the walleyes are feeding, and the crayfish are moving in silence among the rocks. Insects are abroad on the moist, still air. A beaver swims by. Otters dive and play down the channel. Standing on the point, awash in the primeval music of the frogs, I sense something I've caught briefly before in wild and lonely places, but never so strongly. I suddenly sense the existence of a different universe from the one I inhabit. The universe of the frogs and of all these creatures around me. For a moment, under the moon and within the music and burningly aware of all that surrounds me, I know the other universe. Their universe. *The* universe. The universe as it is for countless species upon our planet, as it exists across billions of light-years. A wild, instinctive, primordial universe, of smell and sense and sound, of primitive urges and existential imperatives. For a moment I can feel it. I know its presence and my presence in it.

Suddenly a motorboat roars through the channel and breaks the spell. It passes quickly and the music of the frogs returns, but the feeling—the sense of the other universe—is gone, and, try as I might, I can't recapture it.

But I am left with its residue, its fragrance in my nostrils, its sound in my ears, the feeling of it beneath my feet. I know that I'll long remember listening to the universe of the frogs, that it will become part of a small collection of vivid experiences and memories that tinge a life with new color, imbue everything with overtones of meaning.

I remember a canoe trip when I was feeling low, burdened by worries and a strange sense of disconnection. Even the lakes I paddled and the shorelines I passed seemed somehow far away. The woods were strangely devoid of wildlife, the sorts of encounters I'd so often had. One evening, walking beside a broad beach near my camp in a grove of tall trees, I came to an old white pine, its massive roots reaching over the sandy bank. Gazing at the roots and great trunk, I thought of how much I had always loved the pines of the North and all they'd meant to me.

I vaguely noticed the wind stir in the top of the pine and realized it had been stirring in the treetops all day, somehow just beyond my attention. I sat down against the old pine in the sunset. In the warmth and the red glow. I remember the feeling of asking a question, though I formed no words. After a time a red squirrel chirred and squeaked in the limbs overhead, leaping from one branch to another, stamping his feet, flashing his tail, impatient with me to move along and find some other tree. Eventually he became quiet and silence descended again—just the wind in the tops of the pines. An aspen leaf landed on my hand. More silence, more red and purple stains upon the sky. A dragonfly buzzed close, then alighted on my boot. The smell of dry pine needles was suddenly very strong. The breeze stirred again and rustled in a sapling aspen that I hadn't seen, just beside me.

I sat longer, until the colors faded, and I arose with an unnamable feeling. The feeling stayed with me all evening, even as I fell asleep in the tent, the wind still moving in the tops of the great pines.

When I awoke the next morning, the strange, pleasant mood was still with me. I went about breaking camp with the sense that there was an inner smile somewhere behind my face. I paddled out through mist latticed with sunbeams and headed up a small river. Two loons allowed me to come unusually close. I could see the bright red of their eyes and the softness of their feathers. A beaver swam near and climbed up the mud bank. A bald eagle circled overhead. A mink prowled along the shoreline. I rounded a bend in the river and suddenly—stunningly—there was a bull moose standing in the shallows, big as a building, bathed in misty sunbeams, water lily tubers dripping from his mouth. He slowly sloshed away and disappeared into the willows. He was, I realized, the first moose I had seen in . . . a very long time.

All through the morning paddle the unnamed feeling grew stronger with each encounter, each experience, permeating everything. Until I remembered the wordless question of the night before and sensed that in everything around me was the wordless answer.

Such experiences can be had by anyone. They are not common, but they are not uncommon either. Yet in the context of an individual life, they are golden and transcendent, as though a veil has been lifted and we are seeing the world for the first time. At such moments we are no longer alone and isolated. We are a part of the whole. We are at home in the universe. Writers and thinkers from William James to Abraham Maslow have explored the meaning of such "peak experiences," noting that in their wake one is often left with a profound sense of gratitude, a renewed sense of meaning and purpose.

But it is the strong feeling of belonging to a greater and more mysterious Reality that is perhaps most important. It is this search for connection, or reconnection, that is at the heart of religion itself. The root word of religion—*religio*—means "to bind" or "to connect," and connecting with the Greater Reality, whether we call it God or Brahma or Tao or Mystery or nothing at all, is the essence of life's search.

And so, on this evening, on a granite point under stars and a setting moon, I linger a bit longer. And listen. And know that I'll remember. For I have heard the universe of the frogs. And for a few brief moments, at least, it was my universe as well.

EVERY DAY IN A BEAUTIFUL PLACE is a beautiful day. It's something I've often said as a guide, a dad, an island dweller. Is it true? If it seems occasionally to be not true, is it the fault of the day or the place or the person in the day and place? If I do not hear the singing of the winter wren, is it because he is not singing? If I don't see the sparkle of a wave or the glistening of pine needles, is it the sparkle or glisten that is lacking or something in my eye to see it? If there is a day that I don't smell the perfume of crushed needles underfoot or the living fragrance of the lake itself, is it the earth or the lake that is missing or simply my awareness of it?

There are good reasons to miss these things. Many of these reasons fall under the general heading of worries. And these worries are almost always about something else, someplace else, sometime else. Something about which I can probably do nothing *right now*. That's why I'm *worrying*.

Like the many accomplished worriers of my tribe, I have sometimes raised this essentially worthless pastime to something of an art form, devising new and improved means of worrying. Some of these could be classified as passive worrying and essentially involve variations on the time-honored theme of sitting quietly in one place and stewing. Even for someone who enjoys worrying, this can become tiresome after a spell. By far the more popular, more up-to-date form of worrying is active— or even hyperactive—worrying. This is based on the creative, imaginative notion that if I do something— particularly something stupid, inefficient, or counter-productive—and simply do it *fast enough*, I will somehow remedy the object of my worries. Though there seems to be no objective truth underlying this assumption, its charm is seductive, if not addictive. It leads to lots of circular commotion without initiating much real change. Washing machines work hard, too.

There are other good reasons to miss things as well, other factors that cloud inward skies and still the music in the air. Sometimes it simply boils down to desires— for something else, someplace else, sometime else. Or perhaps regrets—again for something else, someplace else, sometime else.

But in the midst of worries, regrets, and desires, the world is happening. And in the real world, a day in a beautiful place is a beautiful day.

THE OTHER NIGHT I WAS RETURNING from a starlight trip to the Church of Peace when I was startled by a rustling in the junipers. Peering through the darkness I eventually made out the source of the sound—a vague silhouette gingerly moving toward me, standing and wavering like a reed in the wind, then dancing back a few steps before coming forward again. A bit unnerved, I moved quickly toward the cabin where I grabbed a flashlight. The visitor was now snuffling around the wood-splitting area where I'd thrown out a few stale crusts of bread earlier in the day. I pointed the light and the beam pierced the darkness. A pair of golden eyes, a handsome, intelligent face, bushy tail. Fox! His back and tail were dry, legs and underside soaking wet. He must have swum across from another island, coming ashore while I was in the outhouse.

I began to speak to my guest—gently, soothingly, the way I talk to our cat, Tigger. To my amazement the fox came over and laid down on the granite beside the porch, resting its head on its paws. I've often "talked" to wild animals—croaking at ravens, "psssh-pssshing" to chickadees, yodeling at loons, and hooting at owls, but this was obviously a whole new level of successful communication. I know how to talk to foxes!

If he likes to be talked to like a cat, I flawlessly reasoned, perhaps he'd like some cat food. I went to the kitchen and found a leftover can of Tigger's Fancy Feast. I scooped it onto a small plate and set it just outside the screen door, all the while caressing the fox with soft, soothing words. Warily but quickly the fox came forward and, after a moment's hesitation, lowered its head and began to eat, daintily curling its upper lip to reveal immaculate white canines as he took each morsel of "Salmon Dinner." After finishing he retreated a few steps to the rock and laid his head once more on his paws.

"I'm sorry, I'm out of Fancy Feast," I apologized. "Hey, how about some Meow Mix?" Back to the kitchen I went and returned with a bowl of "crunchies." The fox had disappeared into the darkness. I called, and immediately he reappeared. Meow Mix was clearly not in the same league as Fancy Feast as a gourmet dish, but he ate every bit. I went to the kitchen and filled the bowl once more and came back to find that my guest had again vanished. I called softly as I had before, and again he emerged out of the shadows. Once, twice more we repeated this routine, the fox obviously enjoying the unexpected banquet and I delighted with the opportunity to observe such exquisite manners and graceful beauty at close range.

All the time I talked in my best fox-taming voice— no quick or frightening movements, studiously non-threatening postures. With each passing minute, my bushy-tailed friend grew more comfortable in my presence. Clearly, I had a way with foxes that I had never suspected.

When the bowl was empty for the last time, my diner disappeared into the night, and though I called a few more times and whistled and clucked my tongue, he did not come back. Still, it was a marvelous, mystical experience, and I went to bed that night thrilled with my newfound ability to charm foxes.

Today I stopped by Ober's Island and visited with Jean Replinger over a slice of pie and a cup of coffee. I mentioned—quite modestly—my magical encounter with the fox. "Oh, that's neat, Doug. Yep, we've been feeding him all summer. I hear he's been visiting the cabins over on Jackfish, too. He's certainly not starving!"

Hmm. He's certainly not. Hasn't been back to Fawn Island since, either. Probably found someone with a better grade of cat food. But not with a more soothing voice. Of that I'm sure.

I loaf and invite my soul.

Walt Whitman

THE GREAT DANGER of being an island owner or old cabin fixer-upper or Responsible Person of any persuasion is that you can become so busy fixing and improving and taking-care-of that you almost forget to look. And see what's already there. And that it's already good. And probably interesting. And largely unknown to you.

Today I was peeling logs with a drawknife when I noticed a yellow spider. Being busy with Important Work, I gave it only a glance. It is no unusual thing to see a spider on the island. We seem to have many varieties and sometimes even chance upon the fearsome Wolf Spider, which appears capable of eating small dogs. (Kathy, however, has a soft spot for spiders, and any that are found in the cabin are simply covered with a cup, a piece of paper slid under the open end, and the spider returned to the wilds.)

Now this spider, the Little Yellow One, was clearly no dog eater. Yet something about it stuck in my mind, and after a minute or two I went back to look at it more closely. It had moved from the peeled log, where it was a bright yellow, onto the denim shirt I had tossed over the log, where it was a brighter yellow. Day-Glo yellow. Less than a dime in circumference, it had red stripes down either side of its abdomen. On one side of its body it had only two short legs. On the other it had a matching pair, but also a pair two and a half times as long that, whenever I drew close, it waved around in the air. Defensively. Or aggressively. Or both. Or neither. I wasn't sure.

I immediately had the urge to run to the cabin, grab a book, and look up what sort of spider this strange little creature was. But running to the cabin for a book would just amount to another form of improving something—in this case, my mind. While this is an endless and supposedly worthy project, for some reason I decided against it. I decided against improvement and in favor of just seeing. Seeing something that was right in front of me. Seeing something that I knew nothing about.

It can be a bit uncomfortable just sitting with something you don't know or understand. But if you sit with it long enough, it ceases being uncomfortable and is just There. And after it is There for awhile and you are There with it, it becomes something else again. It becomes boring.

Boredom, as an experience, is very underrated. It is, in fact, a great gift. In an age in which time passes much too quickly, it is one of the only ways we really have of slowing down the clock, of truly extending the amount of time we get to spend upon the earth. To be still, doing absolutely nothing for an hour, is to relearn how long an hour really is. Or a day. Imagine a week. We can, I believe, live longer—without actually surviving a single extra instant—simply by giving up our morbid fear of boredom, our need to do battle with it, to fill every moment of every day with Entertainment and Excitement and Accomplishment. And Busyness. Don't bother to ask anyone how they are these days, the answer is always the same. Busy. We say it wearily but proudly, wearing our fatigue as a badge of honor. Not to be busy is to run the risk of feeling unsuccessful, or worse, *appearing* unsuccessful. And to be bored, even for a moment, in the era of hundreds of cable TV channels and Internet and e-mail and supersonic travel and multiple-career households and vitamin-fortified Busyness is, well, almost tragic.

I thought of these things as I watched the yellow spider, feeling bored. Actually appreciating the boredom—the unfilled, unregimented time to be bored. Appreciating a warm summer day—of which, however many I get, I will certainly get too few. And appreciating the spider, sharing the sunlight with me, catching and holding and reflecting it in its bright yellow body.

Finally, after the feeling of appreciation came one more feeling—a rare and unexpected one. The experience of simply being alive. And it had nothing to do with being busy. And I didn't have to look it up in a book. And I didn't have to improve it.

*It is nature's cry to homeless, far-wandering, insatiable man:
"Do not forget your brethren, nor the green wood from which
you sprang. To do so is to invite disaster."*

Loren Eiseley,
The Unexpected Universe

*Nature, Mr. Allnut, is what we are put
on this earth to rise above.*

Katharine Hepburn to Humphrey Bogart,
in *The African Queen*

THERE ARE TWO BASIC WAYS to behold and encounter the natural world. One is to hold it at a mental
distance—to use it, ignore it, admire it, fear it; the other
is to identify with it. The first mode is predicated upon
seeing nature as, essentially, stuff. Pretty stuff, picturesque stuff, dirty stuff, scary stuff, valuable stuff, worthless stuff, living stuff, dead stuff, itchy stuff, smelly stuff.
But stuff. The second mode sees nature as a mysterious,
lyrical realm, pregnant with meaning, inherently instructive and metaphorical to us, who are a part of it.

In the first view, nature is full of "its." In the second,
it is populated by "thous." The first approach is very
popular and has been, by many measures, hugely successful. The second approach is no longer very popular.
In fact, perhaps the only good reason for taking the second view is the quaint and idiosyncratic notion that
we might actually Belong Here! That we did not arrive
from some other place or time or dimension, are not

superior beings left here by some sort of Cosmic Mistake, or grown so smart (*homo sapiens* = "wise man") and advanced that we live only inside our own heads and among artifacts we've created.

Nature is what we are here to rise above.

Baloney.

I'm not much of a believer in either-or choices, having found most of them to be false constructs foisted upon us innocents to fulfill someone else's agenda. But there are rare cases when the choice is real, when things cannot be both/and but must be either/or. And this is one. A big one. At rock bottom, either we belong here or we don't. Either the smell of crushed pine needles, forest breezes, sweet clover, rain, and moist earth is the smell of home or it is not. Either still waters painted with the fires of sunset are beautiful or they are not. The whistling of wings and the high barking of geese is either music or just noise. Either we identify with this world of tree and star, flower and desert, stone and water and belong to it or not. Period. If not, we are essentially aliens, refugees on our own planet.

Our basic task as human beings, our ultimate endeavor, is to demolish our own loneliness. To do this we need the help not only of other men and women, but of trees. And flowers. And animals. Without this help, this company, we are lost upon the earth no matter where we are, lost inside our own skins. Who knows how many psychological vultures with Latin names pick at the insides of people who somehow feel they simply don't belong here? Here at the turn of the millennium, with modern maps, good roads, instant communication, and satellite global positioning systems, it is perhaps easier than ever before to become lost on the earth. It isn't that we don't know where we are, it's that we don't know who we are. And who we are is beings *sharing* with other beings the mystery, terror, and beauty of life. The Lakota have a phrase for it: Mitakuye Oyasin. All My Relations. Traditionally, it is a statement made at any significant or ceremonial occasion, whenever people are gathered—usually in a circle, an inclusive form. It could be made with arms outstretched, an inclusive and vulnerable gesture. It can be made privately, in the silence of one's own heart.

All my relations. The great Christian missionary Albert Schweitzer lived by a motto that carried the same spirit: "Reverence for life." No subject, no verb, nothing that must be done, no one to do it. Just a succinct, sincere acknowledgement. Of belonging. Other traditions have expressed the same insight. From Buddhism: "One nature, perfect and pervading, circulates in all natures." From Hinduism: "In the light of understanding you will see the entire creation within your own soul." From Taoism: "Do not ask whether the great Principle is in this or that. It is in all beings." It does not matter from what culture the insight comes; it is the idea itself that is important. It is an idea that does not grow old or out-of-date, that can be explored as deeply as one cares to take it—poetically, scientifically, spiritually.

This does not mean that there is no conflict, no self-defense, but that conflict and self-defense are managed out of this perspective. It doesn't mean we shouldn't hunt or gather or fish or eat, for life feeding on life is the order of things. Is the order of things "wrong"? It's real, and when it is truly understood and *one feels a part of it*, then empathy and generosity of spirit are deepened.

And that's it, really. Relationship. Kinship. It's not

sophisticated or complicated, only essential. Surrounded by the fragrance of the forest, the stillness of rocks, the movement of birds and wind and stars and planets, it seems obvious—clear as day—that everything has a lesson, a suggestion. Everything that hums or buzzes or flies or crawls or blooms or grows or stands still can be a teacher and a teaching, a part of any decent education. A relative.

To think of the old cedar beside the island trail not as "a tree" but as Grandmother Cedar is to somehow paint the world in warmer, brighter colors, to actually transform the world through the sympathy and imagination through which I perceive it. And then to walk around in it. To belong in it.

I don't have to do this, of course. I can make the other choice, the more conventional choice, and walk around, lost, in a world full of stuff. But in that world people behave as if they have no relatives. And the stuff has no meaning.

And neither do I.

If the sun or moon should doubt,
They'd immediately go out.

Alexander Pope

Although the world is full of suffering, it is also
full of the overcoming of it.

Helen Keller

IN MY TRAVELS I meet a great many people—serious, caring people—who ask me, essentially, how I can be an optimist in this world. There was a time when I wasn't. I'm still not, always. Nobody can be honest and be an optimist all the time. On my good days I am. Most days I am. On the island I usually am. On the island it's always a bit easier to remember why.

Years ago, when I was struggling with my worldview and my self-view, I had a balky car that needed a lot of work. I didn't like getting the car fixed all the time, but I did somehow like the time I spent with the auto mechanic, Don, in his garage. One day I realized why. I was standing around, waiting for my car to get fixed, when another customer came in. He stood around for a while, too. There seemed often to be people standing around in Don's garage. As usual Don engaged in conversation while he twisted wrenches and calibrated things and slid under and out from under cars. At some point the other man said, "Don, you always seem to have a smile or a story or something pleasant to say. You always seem to be in a good mood. How do you do that?" From under a car hood and without missing a beat, Don answered, "Well, I tried it the other way and that didn't work."

I thought it was one of the most profound things I'd ever heard. Simple. Self-evident. A mechanic, by definition, figures out what works and what doesn't work. If something doesn't work, quit doing it. Period. I took Don's words home with me and wrote them in my journal. But I needn't have bothered; I never forgot them. Years later I heard an American Indian teacher say that the only philosophy he was interested in was one that would grow corn. Same idea. Simple. Practical. To the point.

When I was very young, before I got smart and confused, I think I understood this instinctively. Every summer when the clan gathered at Kabetogama, we had a fishing derby. Score was kept diligently, every fish of every size and species noted in the column of the person who caught it. There was great drama and suspense in this contest, at least the way my grandad ran it. But every summer, I was sure of one thing: I would win the fishing contest.

And I did. This certainty that I would win the contest was not an assumption, in which one sits back and waits for something to happen; but a belief through which one makes something happen. So strongly did I believe that I would win the derby, catching more fish than parents, grandparents, aunts, uncles, brothers, and cousins, and so excited was I about the prospect, that I awoke every morning energized and inspired. And, of course, since it was a belief and not an assumption, I did everything possible to make sure that it happened. I ate fast and ran back to the dock and fished while other people were still chewing. I got up early and fished the shoreline while others were still sleeping. I begged and pleaded and cajoled to go out in every boat at any time of day. I read fishing books and kept my hooks sharp and my lures painted, as my grandad taught me.

Looking back, I realize I had stumbled upon a principle that would take me many years to lose, rediscover, and finally be able to put into words: that optimism and pessimism are not opposite ways of looking passively at a static reality—a glass half full or half empty—but instead are opposite means or paths toward creating a *dynamic* reality. Experts and Cynics and Wise Men may speak dismissively of optimism and Pollyanna-ism and rose-tinted glasses and naïveté, and talk in somber tones about "realism"—with the implicit or explicit assumption that realism and pessimism are closely related. But the fact is that in an evolving, dynamic, creative, and circular universe, *optimism is realism.*

To begin in optimism and then act through it is to engage in creating a more optimal or optimistic reality, which circles back again. What you see *is* what you get. And you're ready, you're prepared when opportunities arise and good things happen. If I were to draw a diagram of it I'd simply make an endless, circling arrow with Realism printed at one pole and Optimism at the other.

Of course, pessimism is realism, too, and works by precisely the same principle in the opposite direction.

But pessimism makes me tired. Eventually it makes me depressed. And when I am tired and depressed I'm not worth squat. No fishing, no fish. "I tried it the other way and that didn't work." Right.

There is also, simply, fear. Pessimism (and its kissin' cousin irony—an attitude of smug and distant cynicism that passes for sophistication in our day) is often just fear in fancy clothes, in intellectual disguise. Plain old fear of being hurt. Fear of failing. Fear of loss. Fear of wasted effort. Even the fear of appearing foolish or naive. (It took me a long time to learn—to relearn—not to be afraid to be naive.) "Protected" by fear, one dares not risk optimism or even hope. And every day life gets a little smaller.

Finally, in a complicated and confusing time (insert any dates), as a member of an imperfect species on a beautiful but degraded planet, I can choose to take a buoyant view of the human prospect because I believe that something, perhaps something we call the human spirit, is "smarter" than the human mind. It is also braver and more persistent. It knows that life is often full of pathos and tragedy, nullity and confusion, but it senses something more—a fundamental reality of numinous meaning, in which life is rooted and toward which it grows and aspires. It knows, this something deep inside our deepest insides, what it needs, and its needs are simple. A green world. Clear, reflective waters. Sweet air. Belonging. Justice. Beauty. Hope. Love. A few other things. And although human beings are incredibly adaptable, the human spirit will not endlessly tolerate the lack of its essential needs. Our history— over the long view—proves it. Like a green plant reaching for the sun or a rhizome that grows and flowers again and again, the spirit will keep on. It will persist and find a way.

It is not our minds, clouded by pessimism or doubt, but our spirits that will prevail. And so, we are allowed—required—to be optimistic.

They said it couldn't be done, but that doesn't always work.

Casey Stengel

The first peace, which is the most important, is that which comes within the souls of people when they realize their relationship, their oneness, with the universe and all its Powers, and when they realize that at the center of the universe dwells Wakan-Tanka, and that this center is really everywhere, it is within each of us.

Black Elk

Be not afraid of the universe.

Eskimo saying

BUT IS THE UNIVERSE A FRIENDLY PLACE? It's the question I lobbed at Bill Holm that late night on Mallard Island. "If you think," said Stefansson. "If you think," answered Bill Holm.

Approximately fifteen billion years ago—give or take a billion—in a time when there was no time, in a place where there was no place, and therefore no time or place for anything to happen, something happened. A microscopic pinpoint of condensed possibility exploded, hurling an entire cosmos outward at incomprehensible speed. How did this happen? We don't know. Why did it happen? We *really* don't know. Sometime after that explosion, maybe half a billion years later, stars began to form and the universe lit up.

We now live in that illuminated universe. We might not have. The cosmos released in that stupendous explosion bears many remarkable characteristics, but perhaps none more so than this: proportion, degree. Balance. If the Big Bang had been slightly more violent or gravity imperceptibly weaker, the universe would have dispersed into an insipid brew of icy dust clouds, too thin to coalesce into stars and planets. And rock tripe. And human beings. Had the explosion been slightly less violent and gravity slightly stronger, the incipient universe would have collapsed back upon itself too quickly for the expanding cosmos to have formed. A few minutes, perhaps, a few million years . . . Moreover, in the cataclysmic battle of annihilation between matter and antimatter that ensued in the first few seconds after the original moment, things were eventually decided in favor of a residue of matter, by a ratio of about one part per one hundred million. Or one-millionth of one percent. And there is the tricky question of our universe's natural laws themselves, which could have evolved into slightly different constants in which only the lightest elements would have formed. No carbon, oxygen, minerals—no organic life. In other words, science suggests that the odds against the universe forming as it has and us emerging from it were of a magnitude that defies description.

So says the current science, the current thinking. At least some of it. What does this mean? Well, I don't know. I count on my fingers and sometimes have trouble starting boats, and I just don't know. But it might suggest that something about the universe, or something about what caused the universe, is in favor of . . . life. Or at least stars and warmth and light. And planets that occasionally orbit about stars at precisely the right distance and spin at just the right speed so that life might emerge. It would seem that something wants jack pines. And maybe Jesus and Lao-tzu. Mozart and Shakespeare. Darwin and Einstein. It would seem that the universe

desires life or it wouldn't have called it forth. That perhaps the balance favors us.

For ages previous to ours, humankind has been deeply concerned with the concept of balance, of walking in balance in a delicately balanced cosmos. Today that concept is in eclipse, secondary to such concerns as Going Fast and Getting More. If you can keep your balance, fine, but there are more important things.

No, there aren't.

In an inconceivably vast universe (forty billion galaxies and four thousand billion billion stars are the most conservative estimates), a universe that seems to welcome life, what would be an appropriate response on the part of . . . human beings, for instance? Well, respect certainly. Even awe. Curiosity. And gratitude—for life. A part of *keeping our balance* lies in maintaining and nurturing the capacity for gratitude. When that capacity is lost, bad things happen. Personally. Globally.

The universe is a friendly place, if you think.

I think . . . not. As much as I value and enjoy thinking—about trees and birds and people and literature and science and the planet and the origin and fate of the universe, and other stuff—I suspect that thinking is not enough. I suspect that the universe—or an island or any other particular part of it—becomes a friendly place only when a person *feels*—feeling balance, feeling simple gratitude, feeling the experience of being a live human being standing upon the earth.

When we are in tune with such feelings, Reality itself seems marked by a sort of cosmic generosity. In the nearsightedness of mundane self-interest, the cosmos does not often *seem* generous. We human beings are endlessly thwarted and bedeviled by obstacles big and small that prevent us from getting what we want, when we want it.

But with mindfulness and gratitude, immersion in things green and natural and real, the perspective of open horizons, there are moments when the walls of ego and gloomy self-concern are shattered and the universe is illuminated, just as it was fourteen and a half billion years ago. Suddenly the beauty of a sunrise is a surprise, moonlight sparkling on wavelets is a diamond path, the scent of a wild rose is more delicious than should be possible—or, perhaps a better word, necessary—and life itself is an undeserved and unimaginable bequest. At such times all is grace, and a person may literally "come to himself," come into the heritage of his or her own life. And into the heritage of a friendly universe.

For in the final analysis, each of us dwells within our own inner portion of outer space, living more in the universe we make than in the one we find. As Origen said nearly two thousand years ago, "Thou art a second world in miniature. The sun and moon are within thee, and also the stars." The path toward discovering the universe—yours or mine or Black Elk's or Einstein's—has never stretched out in linear miles or even light-years. It is found in the imagination and reverence with which we regard our world, the actions with which we honor it. And its friendliness is truly up to us.

I arise today
Through the strength of heaven:
Light of sun,
Radiance of moon,
Splendor of fire.
Speed of lightning,
Swiftness of wind,
Depth of sea,
Stability of earth,
Firmness of rock.

Saint Patrick, fifth century

And the round ocean and the living air,
And the blue sky, and in the mind of man;
A motion and a spirit, that impels
All thinking things, all objects of all thought;
And rolls through all things.

William Wordsworth,
"Tintern Abbey"

When you arise in the morning,
give thanks for the morning light,
for your life and strength.
Give thanks for your food
and the joy of living.
If you see no reason for giving thanks,
the fault is in yourself.

Tecumseh

FOR A LONG TIME, Fawn Island was a longing, a dream. As a dream it hovered in the gauzy mists of perfection, its image haunting me, challenging and reassuring at the same time. I remember vividly the day the dream became reality—the wonderful feeling of excitement and anticipation that was mine as we docked at the island. I remember prying the old "For Sale" sign off the cabin. I remember standing high on the ledge, looking out over the lake and saying silently to my childhood hero, "I made it, Grandad," and noticing the distant islands becoming blurred and misty for a few moments. I can recall lying in the sleeping porch one afternoon that first summer, the shutters open, listening to the birch leaves shivering and to the boys running up and down the stone steps and playing on the dock, and the feeling of peace that swept over me. And I remember the joy of discovering each new secret the island and the old cabin had to offer.

But something was also lost when the dream came true—the dream itself. For the only way to keep a dream intact, in all its inviolate perfection, is to never live it out, never realize it. The day a dream becomes real it becomes subject to all the imperfections of reality. The realization of any dream, particularly those things that bring deepest joy—love, marriage, the birth of a child, reaching a lifelong goal—also brings care and worry, the risk of loss, the awareness of impermanence. The surest way to end a dream, in fact, is to reach it; the only way to keep it, to never live it out.

So how does a person who believes in the pursuit of dreams—in the possibility of attaining them—deal with this paradox? By consciously abandoning the quest for perfection in any material sense. According to the old storyteller of Genesis, even the Lord Himself, in creating the universe, did not achieve, or intend, perfection. He pronounced creation not perfect but "very good." The Navajo and other American tribes, when creating a mandala or sand painting, always leave a small opening in the circle, leave something out, or incorporate some small imperfection, so as not to offend the Powers or overstep their bounds. In the realm of time and space and matter, once we are in the company of "this" there is also "that." Up brings down, before creates after, gain is balanced by loss, joy accompanied by sorrow. A dream need not be subject to these limitations. A reality must be—even an island.

So Fawn Island is not perfect. It is not the Kabetogama of my childhood or of my idealized memories of childhood. Which, I suppose, is part of what I was after. Too many people are missing whom I would love to have shared it with; too much cannot be recaptured. Droughts sometimes come and the beetles eat the pines. Storms knock them over. There's occasionally too much noise and boat traffic. Of course, our boat is someone else's traffic. The island is not a pure wilderness hideaway, as wild and remote as I would like. And if it were, we could not reach it with any frequency. Like most folks, I fret over finances and wonder if the Fawn will always be ours. The island, unlike the dream, is not quite perfect.

But it is "very good." And one of the things most good about the realization of a dream—any dream— is that it lends courage and conviction to the future, to the dreaming of new dreams. Some will be realized, some will not; none will be perfect, but all will be worth the dreaming. Here on the Fawn I am close to the past— a past of geology and evolution, tribal traditions and voyageurs, lumbering and resorts, and decades of family memories. But part of the balance of "very good" lies in anticipating and welcoming the beckoning future, the place where all dreams live, along with the opportunity to make them real and share them.

THE OLD CRESTLINER TUGS GENTLY but insistently at her lines. All is stowed and tucked away, the cabin locked up. The song sparrows and whitethroats still sing their morning arias, but the sun is high. It's past time to go.

I always linger when it's time to leave the Fawn. But I know I'll soon be back, and the island will rollick with voices and laughter, the clink of blueberry buckets, the splashing of swimmers, the slamming of screen doors, with precious little time for solitary wondering. And it will be very good.

I pull away from the dock and turn the bow toward Bald Rock. A hundred yards out I turn and look once more. The boulder stands guard on the point, the cabin nestles into the pines. A gull spins lazy circles in a too-blue sky, over a small bit of rock and tree that is the center of gravity around which a few lives orbit each

summer. A young one once told me that Fawn Island is a "lucky place"—perhaps the best description of all. And we who come to walk its trails, gather its fruits, watch its sunsets, and listen to its music know that we are lucky to spend some of the best days of our lives here.

DOUGLAS WOOD is the best-selling author of thirty-five books, including *Deep Woods, Wild Waters* and *Paddle Whispers,* both published by the University of Minnesota Press. His first book, *Old Turtle,* was awarded Book of the Year by the American Booksellers' Association and by the International Reading Association. His other honors include the Christopher Medal, Parents' Choice Award, Smithsonian Notable Book Award, and the Lifetime Achievement Award from the Minnesota Association for Environmental Education. He lives with his family in a log cabin by the Upper Mississippi River.